# WWII MORTARMAN

By Tony Gentry

*WWII Mortarman*

*The Coal Tower* - novel

*Last Rites* - stories

*Yearnful Raves* - poems

Young Adult Biographies:

*Paul Laurence Dunbar*

*Jesse Owens*

*Dizzy Gillespie*

*Alice Walker*

*Elvis Presley*

# WWII MORTARMAN

*My Father's Service with the  
99th Chemical Mortar Battalion:  
European Theater*

**Tony Gentry**

NExTExT Books  
Richmond, VA

Copyright © 2024 Tony Gentry

Published by NExT, LLC
Richmond, VA

All rights reserved.

ISBN: 978-1-7327608-4-4

Cover photograph:
Braun, Mortar Crew at Rhine Beachhead, 1945.
U.S. Office of War Information
(public domain)
http://www.defenseimagery.mil/imagery.
html#guid=5132f9b36dc2320d79f1ae28c0f084839

*For Daddy's descendants, who like me*

*wouldn't be here if he hadn't made it back*

# Preface

Mama always told us not to ask Daddy anything about the War, but growing up in the 1960s, that was not easy. TV shows at the time included *Hogan's Heroes* (a comedy improbably set in a Nazi Prisoner of War camp), *Combat* (a harrowing World War II drama series), *12 O'Clock High* (about a B-17 bomber battalion) and my favorite *The Rat Patrol* (in which a scruffy crew in a machine-gun armed jeep face off against Panzer tanks in the North African desert). For years I avidly read the monthly comics *Sgt. Rock* and *Sgt. Fury and his Howling Commandos* (the latter starring Nick Fury, who went on to lead the spy agency SHIELD in middle age), along with the weird *GI Combat* series, about an American tank crew haunted (and protected by) the ghost of Confederate General W.E.B. Stuart. Wearing our plastic Army helmets and canteen belts, brother Greg and I fired replica Tommy guns in endless neighborhood war games with friends, squirming along the ground on our elbows like the G.I.'s on tv, lobbing stubby corn stalk roots that served for grenades (the Nazi potato masher versions), and boarding the ladders of side-railed coal cars that we imagined to be tanks or landing vessels. What Daddy thought of our romping about aping an era he surely wished to forget I can't say.

Our house held tantalizing evidence of his wartime service, a handful of ripple-edged black-and-white photographs from the earliest days of Mama and Daddy's marriage, she giddy in the blowsy skirt and saddle oxfords women wore in the 1940s, he dashing in his uniform. There were even some shots he'd sent his bride from overseas. These pictures were jumbled among others in a dining room cabinet, where on rainy days we four kids could dig them out and marvel that our parents had once been young. I've included all the wartime photos that are left in this book.

We had a pair of relics, too. One was an old ornately carved dueling pistol that Daddy left laying out on the fireplace mantle. Greg and I often used it as a toy, including it in our war games, though it was so heavy that we preferred our plastic weapons. Mama finally gave in to our insistent questions and allowed that Daddy had lifted it from a university display case somewhere in Germany at war's end. His other booty was the real treasure, though. The swastika section of a red Nazi flag, raggedly knife-cut to fit his pack and signed in ink by 28 of his buddies, their addresses – Fall River, Mass; Plano, TX; Pocatello, Idaho; Saginaw, Michigan – scrawled blurrily below their names. On the upper right corner, Daddy had inked in the date: March 26, 1945. Mama kept that artifact in a wooden chest in their bedroom, but we could go in any time, it seemed, to rifle through her keepsakes and pull it out to study that devilish crooked cross happily defiled by the neatly penned signatures of Daddy's wartime pals.

That flag fragment hangs framed in my study today (with a caption explaining that, no I'm not a latter-day Nazi). In attempting this narrative, I've tried to track down the men who signed the flag. Too late. The youngest among them would be a centenarian, and nothing has come of my inquiries. Why didn't I make this effort earlier? Why didn't I finally overrule Mama's injunction and ask Daddy about those old pals or try to find them on my own? Well, all I can say is that childhood taboos are hard to break. And it was clear, even in old age, that Daddy preferred not to reflect on wartime memories.

Well, that's not strictly true. Apparently, when chatting with my sister Kay's husband Butch, he'd loosen up and share a few choice anecdotes. Butch died suddenly in 2022, before I'd learned about those talks. So, again, too late. From time to time, though, a brief story would slip out. I overheard him reminiscing with another WWII veteran one day, sharing an eye-opening tale of wartime bliss. And he startled me one night, leaping up to snap off the tv when a guy being interviewed on the old Tom Snyder talk show claimed the Holocaust was a myth. Daddy stood there in his t-shirt and briefs, in a haze of cigarette smoke, and declared: "He's a liar! I've seen bodies stacked from here to Fork Union." (Fork Union is a town a few miles from our home.) Other anecdotes, as I recall them, pepper this narrative.

The rest of Daddy's wartime memories died with him. Mama's gone, too, and all of his six siblings have joined them, either in the Fork Union Baptist Church cemetery or at the old Lyles Church graveyard near their childhood home. What happened, I wonder, to the letters Mama and Daddy mailed each other during the war? I have one postcard that Daddy sent to his older sister Dorothy from North Africa at Christmastime 1943. That's all.

Mama wrote a memoir that recalls her childhood during the Great Depression and what it was like to be a wartime bride. As she tells it, Daddy used to whistle a bob white quail call when he came a-courtin'. And that's how she knew he was home from the war, hearing that call again from her bedroom window. The title of her book *Then the Bob White Called* honors her precious memory. So we do have her side of the story, at least.

What follows is my best effort, such as it is, to tell his. I've tried to patch together Daddy's path through the War, stitching his few personal anecdotes onto sparse documentation about his Army battalion.[1] Over time, if I learn more, I'll thread that in, too. This story is for Daddy's descendants, my sons among them, who might occasionally wonder about their ancestors from the Greatest Generation. It's one story among many millions thrown up by the global catastrophe that was World War II. I did this for me, too, of course. A much-belated effort to come up with some answers to the question I never dared ask: What did you do in the War, Daddy?

---

[1] Fortunately, WWII was heavily documented. I've pulled from a shelf full of pertinent wartime narratives (see the bibliography at the end), and in footnotes, I've offered links to YouTube film clips that do a better job than my story does at evoking that era before tv's, cell phones, computers, and air-conditioning, when the whole country was pulling together for a cause.

# Introduction

So how do you write a book about the wartime experiences of a man who rarely mentioned it? Fortunately, I had his draft card and his discharge certificate, which marked the beginning and end of his service, but also crucially named the battalion and company he served with. That information led to a brief online summary of his unit's path through the war that had been compiled by some of its officers.[2] As far as I know, it's the only published document written specifically by and about the 99th Chemical Mortar Battalion. It became my blueprint and map, and I highly recommend that you read it. That document names all the places where the 99th went (down to the company level), gives the dates and names of the battalion's attachments to anti-aircraft and infantry divisions, and provides a few choice anecdotes along the way. A separate link lists the officers and enlisted men (down to the rank of corporal) who served in the battalion.

Which led me down a productive rabbit hole, because the activities of the other combat divisions listed on the webpage are well documented online, in published histories, and in the case of the 101st Airborne Division,

---

[2]https://www.4point2.org/hist-99.htm

in an entire tv series (HBO's *Band of Brothers*). My research, then, consisted mainly of matching the 99th's outline with the more comprehensively documented histories of the infantry units they served alongside, and I must say that has been a riveting, enlightening and sometimes harrowing experience. Daddy had never spoken about the winter of 1945 in Alsace, the coldest and snowiest on record, which he spent mostly outdoors. He never said anything about Operation Nordwind or the Colmar Pocket or the Maginot and Siegfried Lines or any of the 85 villages and towns that his battalion fought over during that winter. But other people did. And learning about what he and his comrades went through, as I gradually pieced it together, has been both gratifying and painful, and both for the same reason: A long time coming, but I finally have a fair idea of what my father experienced in the war.

This book draws on a few other key sources. The dozen or so extant photographs that Daddy sent home are included, along with photographs of the Nazi flag his buddies signed, the dueling pistol he brought home as booty, and the Christmas card he sent Aunt Dorothy from North Africa (no other correspondence has survived). I've patched in other pertinent pictures and maps from books and the internet, in an effort to clarify some of the battlefield confusion. Most important, I think, are the precious few tales our parents shared, no doubt burnished by my faulty memory, but included here as honestly as I can recall them.

I will keep trying to learn more and will revise the story as things turn up. Sadly, Daddy's detailed service record appears to have been lost in a fire at the Army's data storage library in St. Louis back in the 1970s, along with most of the service records of World War II Army veterans. But there are other repositories scattered about the country that may help flesh out this story, and I hope to visit them as time allows. Or maybe you will pick up the thread and sew up this quilt's loose ends.

Writing this story has led to some personal regret and reflection. If I'd only known what this research has taught me when Daddy was alive; if I'd only pursued it when the pals whose names decorate his flag were around to interview. Yes, but even this incomplete record has helped me understand him better, which means that it also helps me better understand our family and myself. What those millions of old warriors went through when they were boys and girls so long ago has shaped each of us who came after them and all the world we live in. A lot to chew on there.

# Let's Go Back

The names and hometown addresses on that old Nazi flag, signed with swirling, elegant penmanship no longer taught in schools, help in imagining the young men in Daddy's battalion. Like him, they were all hard nuts rounding 20, draftees pulled away from Dust Bowl farms, seafront docks, small town feed stores, and mountain coal mines all fallen prey to the Great Depression. Boys with no prior notion of any geography beyond the march of a mule-drawn wagon or the reach of a jalopy's tank of gas. The rigors of war proved different in scale, but similar in some ways to what they'd known back home. Except, of course, for the killing part. That was new.

Daddy grew up on a share-cropping farm near Palmyra, Virginia, the seat of rural Fluvanna County, as the second youngest of seven kids, four boys and three girls. Our family had been in Fluvanna since the late 1600s. His grandfather had owned a stretch of rich land along the Rivanna River, and apparently he had owned several people, too. Despite that fact, he was thought of, among the White folk, at least, as a God-fearing and righteous gentleman, a co-founder of Effort Baptist Church. He lost that farm after the Civil War, so his sons found themselves tilling and harvesting on the share system for the carpet baggers who took over their land. Daddy recalled visits to his grandparents, when he and Uncle Jack would climb into their attic to rifle through a chest full of worthless paper money printed by the county during the War. Which, of course, is how they'd lost the farm.

Aunt Dorothy, Daddy's older sister, lived within a few miles of her childhood home all her life. She baked the best caramel cake in the world. And one day years ago, while I sat pressing a fork into the last crumbs on my saucer, she surprised me with a memory of the carefree, happy-go-lucky, freckled fellow with the coppery-blond hair who was her younger brother and my dad. She said, "Lyn was never a worker. He never cared for workin'."

*Daddy and his younger brother Mac in farm wagon, older brother Jimmy standing. All three were later called up for the War.*

This about a man who in my experience did little but work, at least 60 hours a week as a butcher at the local supermarket, with home chores squeezed in around the edges. I quizzed her on this, followed up by interviews with Uncle Jack, who was Daddy's big brother and mentor, and came away with a somewhat different take on Aunt Dorothy's notions about her purportedly lazy younger brother.

Which led me to attempt a never-finished lengthy poem that recounts what I'd learned about Daddy's youth. Allow me to share some lines of it here, poor reader, to foreground the wartime story and to hint at a different era, nearly a hundred years ago, when farm life was not so different from the way it had always been. One other point before burdening you with this bedraggled ode. Daddy was pulled out of school after sixth grade to work the farm. From the age of 13, then, he was a full-time working man, doing stuff like this:

> Lyn was never a worker
> he never cared for working
> Aunt Dorothy said
> loved to play
> in that cave dug out of the hill
> sloping down in front of the porch
> down to the spring rocky and clear
> green hill sprigged with rabbit tobacco
> chewed by the boys all healthy sons
> struggling together to make it work
> that slog back up with the water.
>
> Tenant farming
> around the old house
> they called it Hilldale
> corn mostly wheat
> vegetables to can
> some tobacco
> some bushels some bails
> for the man
> while the girls went to school

(all heavy woods now
tracked with lumber trails).

The farm didn't pay so
they got hired out to a mean
old skin-flint woman
Jack & Lyn milking
slopping slaughtering mowing
forking grinding gristing
chopping splitting hewing stacking
shoveling hammering sawing post hole
digging filling sweeping mopping
scrubbing painting patching laughing
joking pissing spitting shitting back of the shed
and she always with an eye out
always fretting who is cheating who is
stealing who is wasting who is loafing
and nothing but applesauce
and fat back for lunch
and fetch me a bucket from the spring.

And then the Academy 60 milk cows
Guernseys bucket on bucket twice a day
that's 30 a piece at 5 in the morning
herd 'em in the stalls fill the feed troughs
and by sunup they were done
hands raw as meat but over time growing
strong and thick tough yet tender
to the tug and again in the afternoon
those bitter winters in the '30s
everywhere glazed for weeks
crystal palace trees blinding
forests stooped and glaring
so you couldn't look
but bent like bejeweled dowery maids all
grotesque and ruined you had to
pull yourself up the fence
hand over hand and slipping.

Bad winters
poor summers
potatoes rabbits
but he was courting then
Jack brought him along to visit his girl Nellie
and there she was her younger sister
(Jack said he acted like he'd never seen a girl before)
Virginia played coy try to be like big sister
but smiled her toothy smile
he was a narrow six feet
broad shoulders strong arms and hands
like a tender vise
shy with an imp smile
that didn't really fret her daddy even
quiet and respectful freckled and ginger-haired
oh yes she liked him his eyes were dreamy.

Jack & Nellie married got a car
Lyn & Virginia borrowed it sometimes
double features at the high school auditorium
he pulled up on a back road once and
sort of touched her knee but
thought better of it immediately.

One winter night all the way to the Byrd Theatre
in Richmond Bing Crosby's *Holiday Inn*
emerged to news boy shouting Pearl Harbor Day
the original they both knew what that meant
day before Valentine's they married
and made a home for half a year
and then the draft card arrived
and then the train.

# The Year Before

Mama's memoir paints an idyllic Spring and Summer of 1942, young newlyweds giddy with love in a sparsely-furnished apartment in sleepy Fork Union, while the newsreels shown before movies at the high school auditorium told of battlefield horrors far away. She, a graduate of that brand new high school, had found a clerical job with the county school board; he stayed on milking cows at the military academy. Though Daddy had turned 21 two weeks after their wedding – draft age – at that time farm workers and married men were still deferred from service. But late in the summer, that changed. The draft age was dropped to 18, along with the farm work and marriage deferments. When the inevitable call-up arrived, Daddy took a bus to Richmond and stood in his briefs with a line of other young men for the Army physical examination. A month later his draft card, postmarked September 22, followed.

Here it is:

Already Daddy's older brother Jimmy had joined the Army and his baby brother Mac would soon be called up, too, along with Mama's brother Hollis (who joined a tank battalion in Patton's 3rd Army and won a Purple Heart at the Battle of the Bulge), and their cousin Curtis (who would be killed somewhere in France).

*Daddy's older brother Jimmy on left;
Mama's younger brother Hollis Glass on right.*

The newsreels that summer told a tale of global war, the world at a tipping point. France had surrendered to the Germans and doughty England had suffered a desperate evacuation of their whole army at Dunkirk, then beat back Luftwaffe air assaults in what the newspapers were calling the Battle of Britain. On the Eastern Front, having divided Poland between them, German and Russian forces clashed. America had just joined the fray, facing off against Panzer tank divisions in North Africa. Newsreels displayed blurry footage of behemoth warships and waspish fighter planes going at it in the Pacific battles of Midway and Guadalcanal. The newsreels were exciting, frightening, confusing. When on Halloween Day Daddy boarded a train in Richmond for basic training further down the James River, he stepped onto the roulette wheel of war, with no idea where fate might take him.

# Fort Eustis, Virginia

Fort Eustis sits just downstream from the nation's cradle Jamestown, at the James River's final bend before it spills into the Chesapeake Bay. Since World War I the fort had served as a field artillery training facility. To recruits, it may have seemed a stroke of luck to do basic training there. Everyone knew that the worst job in the Army was infantry, where you served as cannon fodder. Much better to be an artilleryman, lining up well behind the infantry and firing cannons over their heads onto enemy lines. At least half a step back from chaos, right?

Lyn[3] stepped onto the base into a highly organized soldier assembly line. They swapped out his farm clothes for khaki's, shaved off his coppery curls, hung a dog tag around his neck, and lined him up with two hundred other young men from all over the Eastern seaboard for ten weeks of basic training.[4] On his cot in the barracks, he found a copy of the IRTC Handbook – Fort Eustis, which sought to explain his situation:

---

[3] I'm going to use Daddy's name from here on; it seems more natural somehow, as he was a young man and we kids were not yet even a gleam in his eye.

[4] By 1943, Army basic training typically lasted about seventeen weeks. The first ten weeks or so was boot camp, and the following weeks focused on

> You are here as a trainee because your country is engaged in a life and death struggle with one of the most powerful and ruthless combinations of powers the world has ever known. Our enemies are tough, cruel, and highly-trained. Their defeat is essential before this world can become a decent place in which to live. The defeat of the enemy cannot be accomplished by amateurs or by half-trained men. During your intensive training you will find the work hard, the hours long, the going tough in spots, as it should be. If, however, you concentrate your energy and your determination on your work, you will become an alert, confident, skilled soldier, prepared to function as a member of a combat division.[5]

Fort Eustis spread across a flat and mosquito-infested peninsula, made up of broad stretches of bare fields lined with scraggly pines, all interwoven with muddy trails and sucking swamp. Back home in Fluvanna, you could throw a rock across the James River. Downstream at Fort Eustis, it was more than two miles wide. And there was a nasty kind of snake in the creeks, a fat black serpent that bared a foamy mouth with poisonous fangs. They called it a water moccasin. Just a hundred miles from home, Lyn had landed in a strange and unfamiliar world.

During his first days at Fort Eustis, Lyn lined up for small-pox, typhoid-paratyphoid, tetanus and other vaccines, a humiliating "short arms inspection, and a course on "sex morality". He took aptitude tests for different battle roles, signed up to get his teeth looked at and watched a silly Hollywood movie or two. Then he went to work. The idea of boot camp,

---

specialized training, for roles such as sharpshooters, tank crewmen, radiomen, mine sappers, etc. For Lyn, specialized training as an artilleryman occurred after he left Fort Eustis.

[5] From the memoir *War Makes Men of Boys: A Soldier's World War II*, by Katherine I. Mille.

then as now, was to toughen recruits up physically, to teach them to fight, but most importantly, to have them each learn to act as part of a team instead of as an individual. Following strict commands, they marched, crawled, dug, fired, cleaned rifles, peeled spuds, and climbed walls. The smallest mistakes were punished with kitchen or latrine duty or extra physical training, sometimes shared by the whole battalion.[6]

A typical day of basic training looked like this: Roll call at 5:55 am, allowing trainees 25 minutes to wash, dress, make their bunks, and march to the mess hall for a 20-minute breakfast, then back to barracks to collect their packs and equipment for a 40-minute march to a training site, where for the next nine hours they drilled, with a short break at noon for a canned lunch. At 5:30 they marched back to camp, had supper, a 30-minute allowance for showers, and then housekeeping chores and cleaning of equipment. At 9:45 the lights went out. This made for a rigorous 16-hour day that might involve grueling mandatory 25-mile road marches with full equipment or live fire obstacle courses. Not everybody could take it, but most did, and as the weeks went on, the recruits toughened up, which is exactly what the Army wanted.[7]

Mama's memoir recounts how she moved to Richmond and worked in a DuPont nylon factory that was making bomber tires, so she could be within reach of Lyn's weekend passes. She wrote that they spent every weekend together during basic training. No doubt, the two made a cute couple, he in his fatigues and she in her baby blue Rosey Riveter jumpsuit.

---

[6] Probably the bleakest film rendition of basic training is the Vietnam War epic *Full Metal Jacket*. The silliest may be Bill Murray's *Stripes*. Neither, however, is about WWII boot camp, rather Vietnam War era basic.

[7] Kennett, Lee. *The American Soldier in WWII*, p 53-4.

Here is a photograph of the couple taken in front of Aunt Nellie and Uncle Jack's house. If the caption is correct, then Lyn also made it back home to Fork Union at least once during boot camp.

But then he disappeared, shipped west without notice to Fort Bliss, Texas for artillery training. That journey involved a nearly two-week ride across eight states onboard a troop train, made up of old Pullman passenger cars converted to hold as many men as possible. Sleeping berths were bunks. Two men slept together on the bottom bunk; the guy on the top bunk got to sleep alone, but so close to the ceiling that he couldn't turn over. Air-conditioning was an open window, if the window wasn't stuck closed.

Troop trains had their own kitchens that made box meals as they rolled along. Once a day, the trains stopped and let the men off, all of them filing onto a nearby field for calisthenics and on occasion getting a group shower from the spout of a water tower.[8] The trains idled on siderails for hours at a time, while other trains hauling military gear (tanks, trucks, and armaments) took priority on the main tracks. Fortunately, at some stops, townspeople offered donuts and cookies, or even a whole frosted cake for a few lucky men to share.[9] Troop trains didn't follow a straight line. They meandered along, picking up additional riders at forts along the way. It was a long, crowded, smelly and generally boring way to travel. But for the first time in his life, Lyn was beyond the bounds of the Commonwealth of Virginia, gazing out a grimy window as mile after mile of the American South passed by.

---

[8] 11-minute Military Department film clip titled *Troop Train*: https://www.youtube.com/watch?v=IIrSwQr4A3c.

[9] Mama's memoir recounts a sweet anecdote from this journey.

# Fort Bliss, Texas

At Fort Bliss, Lyn joined the Army battalion that would be his throughout the War. The 442nd AAW (Antiaircraft Weapons) Battalion included roughly 500 men divided into four Companies that were learning to operate 40-millimeter automatic weapons as part of a mobile artillery unit. The battalion had been created at Fort Bliss in October and had been adding men ever since. Arriving in mid-January, Lyn was a late addition.

Fort Bliss, like its neighboring city El Paso, sits at the westernmost point of Texas and occupies an enormous and bleak desert mesa that stretches into New Mexico, hard up against the Rio Grande River that marks the Mexican border. The fort has a storied history dating back to the Indian Wars, but in World War II its mission, like that of Fort Eustis, was antiaircraft artillery training. During the War, its size dramatically expanded from a few thousand acres to the more than a million that it occupies today. When Lyn stepped off the train from Fort Eustis, he joined a quarter of a million other soldiers who trained at Fort Bliss, the largest and most sophisticated anti-aircraft training facility in the world. The fort held dental clinics, a fire house, a library, several chapels, and two huge air-conditioned movie theaters. But most of what it had was sand.

If Lyn thought Fort Eustis was foreign, the terrain at Fort Bliss must have seemed out of this world, a barren, ruddy desertscape that stretched flat as an anvil way out to a horizon framed by jagged black mountains. The air was so dry that your sweat flaked to salt on your skin and your tongue felt like cardboard, then at night it got so cold, you thought you were on the moon. Not a tree anywhere. Sagebrush and tumbleweed. Skittering scorpions and hairy tarantulas. And a whole new catalog of nasty snakes, ten kinds of rattlesnakes among them. But, if you wanted to practice blowing things up, you couldn't find a better place.

Fort Bliss was an all-purpose college of artillery training. Some battalions learned to manage heavy cannons, the 5-ton Howitzers that could throw a high-explosive shell nine miles; others trained on the massive 30-ton 120 mm anti-aircraft guns that could send their 35-pound shells almost vertically into the sky as high as twelve miles. Others learned to fire smaller "tank killer" 105-mm and 75-mm howitzers that could be pulled by a jeep or even a pack animal on the battlefield.

Lyn's battalion trained on the Bofers 40-millimeter automatic gun, a weapon used by all sides during WWII, and called the M1 automatic by U.S. forces.[10] Considered a midweight anti-aircraft cannon, the M1 could be mounted on a trailer pulled by a truck, emplaced onboard a fighting ship, or set atop a half-track armored vehicle. Chrysler won the contract for these guns in the U.S. and built 60,000, some going to the Navy and others to the Army.[11] Teams of seven were needed to operate a single M1, including a

---

[10] WWII Technical Manual for use of 40-mm Automatic Gun M1: https://maritime.org/doc/boforstm252/index.php.

[11] The Navy mounted M1s on virtually every vessel larger than a landing craft.

squad leader, two gunners, two loaders and two ammunition bearers. They practiced under a stop watch, eventually learning to aim and fire their two-pound high-explosive projectiles at up to 160 rounds a minute. They learned how to switch out the 7.5-foot long barrel when it over-heated, took turns at the different roles and practiced operating the gun with a smaller crew, as they might need to do on the battlefield.[12] They trained at night, too, with floodlights sweeping the sky for targets. Those targets were flags towed on 50-foot ropes by B-25 and B-26 bombers manned by woman pilots of the Women Airforce Service Pilots (WASP) group stationed at adjacent Biggs Army Airfield.[13] On the windswept desert, gun crews spent half their time, it seemed, cleaning grit out of their weapon's gears.

While at Camp Eustis, Lyn had used every pass to visit his bride in Richmond. But that was impossible now. Way out on the Southwest borderland, after a multi-state train ride, he must have begun to grasp how big and how varied America really was. The men at Fort Eustis, most of them from Virginia, talked like he did. Most were Baptists, raised like he was to steer clear of dancing, cards and drink. But at Fort Bliss the mix was different. Some boys from way up near the Canadian border spoke through their noses; others from the Gulf states dripped molasses from their

---

[12] WWII film clip showing an Army M1 gun crew shooting down two enemy planes: https://www.youtube.com/watch?v=yCJ8WVk5IhA.
And here's a montage of M1s in war movies:
https://www.youtube.com/watch?v=AdoQDyO_zDo

[13] A WASP pilot remembers her service at Fort Bliss:
https://www.nbcdfw.com/news/local/women-pilots-from-wwii-return-to-texas-to-celebrate-flights/235594/

tongues. The Jersey boys talked fast as auctioneers. And the Baptists, he discovered, were few and far between.

Figure 1 - 40-mm Automatic Gun M1 (AA) and 40-mm Antiaircraft Gun Carriage M2A1 - Firing Position-High Elevation

You had two choices for a weekend pass, the ramshackle town of El Paso or its Mexican sister city Juarez across the narrow Rio Grande. Both towns were made for R&R and both were throwing up Quonset hut bars

---

[14] https://maritime.org/doc/boforstm252/index.php.

and brothels as fast as they could to keep up with the Army base's wartime expansion. Let's imagine a minute:

Lyn had sipped his first beer in Jackson, Mississippi, when the troop train stopped for a night to change engines, but he felt guilty about it. Like he felt guilty about the time Jack had come upon a jar of moonshine cooked by a colored neighbor and they'd drunk themselves sick inside the Lyles Church graveyard walls. Out in the desert, though, the dry air just insisted that you whet your whistle, and when the one El Paso native in the barracks invited everybody out to his uncle's taverna, well, of course Lyn tagged along. The taverna was just a beige adobe box with a clay-shingled roof, but inside was like a church turned upside down. Basically, everything the preacher taught not to do. A bandstand on the wall could hardly hold the ten-piece band with its raging fiddle and pumping accordion; the back tables were packed with card players; couples two-stepped clumsily on the dance floor; and the bar was lined with men in khaki and women in red feeling each other up. It was all noise and smoke and laughter, like nothing he'd ever seen or imagined.

So when Jose from El Paso threw an arm around his shoulder and shoved him towards the bar, he pulled back, abashed. "You ever heard of tequila?" Jose asked.

"To kill ya, what's that?"

"Tequila, yes, it does aim to kill ya, but here, we'll take it slow."

"I don't know. My people don't hold with all this."

"Y'all don't drink?"

"Nor dance nor play cards, neither. None of all this hell-raisin', really."

"Well, Lyn, why is that, may I ask?"

"The Bible says," he replied.

"Oh, that. You ever thought why we're out here shootin' all these guns?"

"Well."

"Point bein', and take this correct, now. I don't mean no harm. Just." He raised a shot glass brimming with a liquid clear as water, his gaze taking in the carnival all about. "What I mean is, the Bible says a lot of things."

Lyn considered that, let his gaze track his friend's, and allowed a slow nod. Hours later, it could have been days or minutes, all the way back to the base, he barfed.

Lyn's training at Fort Bliss lasted only a few weeks, but during that time it must have felt like he had fired a million shells. After all those double-time marches into the hills, the stopwatch timed firing drills at targets in the sky, and the hijinks in El Paso, the men of the 442$^{nd}$ AA battalion were in excellent physical shape, knew their weapons and had grown to see each other as brothers. The newsreels by then showed some good news, the Allies creeping forward in places nobody had ever heard of before. The Americans and the British were ganging up on the Nazi general Rommel's Panzer divisions near Tripoli, the Russians were pummeling the German army at Stalingrad, and the Marines had taken Guadalcanal from the Japanese. Some guys joked that the war would be over before they ever got on a boat, but their sergeants warned that the Nazi's and the Nips were made of sterner stuff than that. Like cornered rats, this is when they'll get up on their hind legs, they said. American army ain't even set foot in Europe yet. You boys will be bloodied before all this is over. You'll see.

# Camp Rucker, Alabama

The men of the 442nd AA Battalion departed Fort Bliss on February 4th, boarding a troop train for the dreary journey across the endless and generally featureless state of Texas into swampy Louisiana and kudzu-draped Mississippi, deboarding in Alabama at a hastily slapped together Camp Rucker for a few weeks of combined training with the 81st Infantry Division, the G.I.'s they'd be partnered with in the field. Squeezed into the far southeastern tip of Alabama, hard up against both Florida and Georgia, the camp was just row on row of canvas-walled huts lined with gravel roads, the land as flat as the Texas desert, but heavily forested in pine and swamp oak. Underfoot lay ruddy red clay, the way dirt was supposed to look. If you squinted, you could almost imagine yourself back home in Fluvanna. No more rattlesnakes, either, though Lyn's ears perked up at a warning about alligators in the lakes thereabouts.

At Camp Rucker the artillerymen began to train side-by-side with their infantry brethren, no longer just shooting at targets trailing tow planes all day, taking double-time marches or digging and refilling holes to keep busy. This was war games. Lyn's Company split off from the rest of the battalion, lined up with a Company from the 81st Infantry Division, and set out through the swampy woods to learn maneuvers, mostly how to find a likely spot to park the truck towing your M1, site the gun and launch so you nailed incoming planes without blowing up your own men, the sky black with "flak" as their shells exploded in a cloud of shrapnel and smoke. End of the day, the infantrymen came back, their uniforms caked with red clay, and everybody bivouacked together in the woods, slapping at mosquitoes in their pup tents.

The artillery crews practiced as riflemen, too, because in the likely event that they'd eventually run out of ammunition in a battle, they might be called on to join an infantry assault.[15] They practiced hand-to-hand combat in case their position was ever overrun. A lot to learn and no time to think hard about any of it. The war was calling.

Lyn spent his first wedding anniversary and his 22nd birthday two weeks later in a swampy Alabama bivouac, his ears ringing from day after day of battlefield simulations. Word was that his battalion would join the 81st Infantry Battalion they'd been training with on a boat to the Pacific theatre to fight the Japanese, but when the 81st left Camp Rucker, Lyn's

---

[15] What a rifleman needed to know, per Army Code 745:
•Loads, aims, and fires a rifle to destroy enemy personnel and to assist in capturing and holding enemy positions.
•Places fire upon designated targets or distributes fire upon portions of enemy line, changing position as situation
demands.
•Must be able to use hand weapons, including rifle, automatic rifle, rocket launcher, rifle grenade launcher, bayonet, trench knife, and hand grenades.
•Must be trained in taking advantage of camouflage, cover and concealment, entrenching, recognition and following of arm and hand signals, and recognition of enemy personnel, vehicles, and aircraft.
•Must be familiar with hand-to-hand fighting techniques. Must understand methods of defense against enemy weapons.

*Photo Mama sent Daddy at some point during his deployment*

AA battalion stayed behind.[16] Then on the last day of March, they climbed aboard drab green Army trucks with canvas tops for the 300-mile ride across Georgia to Camp Stewart, located near Savannah on the Atlantic coast. This was an era before Interstate highways, and the roads through rural Georgia were rutted and narrow, so the going was slow. But people lined up along the streets of small towns and cheered. Some tossed the men peppermints left over from Christmas.

On the trucks, rumors spread. If they weren't headed west to fight the Japanese with their pals in the 81st, then what was the plan? The last newsreel they'd seen at Camp Rucker showed all-night bombing raids by British and American planes over Europe. One raid, the newscaster said, had involved an almost unimaginable 1,000 planes dropping 225 tons of

---

[16] The Wildcats of the 81st won battles on ten South Pacific islands during 166 days of combat, the division earning seven combat awards: https://history.army.mil/html/forcestruc/cbtchron/cc/081id.htm.

bombs on a single French town where the Germans were building U-2 submarines. Another raid had even reached Berlin, as the Germans kept up their own nightly bombing of London and surrounding English cities. But so far, no American had set foot on mainland Europe. What was the ground game? How would the Army join the fray? Where would the men in the convoy fit in?

# Camp Stewart, Georgia

After nearly 20 hours bouncing around in the back of their trucks, all five hundred men in Lyn's battalion unloaded at Camp Stewart in the middle of the night. The air, even in mid-winter, was sultry, and they could smell the funky swamp surrounding them. In the dark, it was like they'd driven in a big circle, landing right back in flat nowheresville Camp Rucker all over again. They filed into makeshift barracks that were basically squared-off canvas tents with pine wood frames alive with hungry mosquitoes for whatever sleep they could get. But when they were mustered into ranks at dawn, things began to sort themselves out. Like Camp Rucker, Camp Stewart was a work in progress, created out of whole cloth for the War as an Anti-Aircraft Artillery Training Center. In order to provide the miles of firing ranges required for this work, thousands of acres in four Georgia counties had been requisitioned for the base, which sat like a squat frog in the middle of a heavily forested swamp.

Things happened fast at Camp Stewart. High overhead, WASP-piloted bombers from nearby Hunter Field towed targets to shoot at. Cannon-fire and ordnance detonations rang out all day. This was where artillery battalions received their final tune-ups before shipping overseas, as wave after wave of men passed through to troop ships waiting at the nearby ports of Savannah and Charleston. Lyn's team was assigned a new name, the 144th Aircraft Artillery Battalion, which yet gave them no clue about where they might be headed.

*Post card:*[17]

Mama's memoir describes her train ride to Savannah in April (her first ever out-of-state journey), to spend a weekend with Daddy. She recounts how she charmed a Military Policeman at the base to go retrieve him from the woods, where his Company was bivouacked, and what she thought of the ruddy mustache he'd grown. The most poignant line of her anecdote: This would be the last time the newlyweds would see each other for nearly three years. It's easy to imagine the young couple wandering the tree-lined streets of Savannah, licking ice cream cones, holding hands,

---

[17] https://www.digitalcommonwealth.org/search/commonwealth:w3763d514.

maybe necking on a park bench in the shade of live oaks draped with gently waving Spanish moss. I think about that mustache. What it must have meant. For one thing, if U.S. Army privates were being allowed to grow facial hair, then that must have been an allowance for the fact that the Camp Stewart thousands were soon headed into the maw of war. Lyn's first mustache would have signified other things to him. He was a trained and ready soldier, who could hit a target six miles away with a 40-mm artillery shell. He could read a poker tell and preferred whiskey to beer. He'd seen the South and had pals from half the states in the U.S. More than all that, he was 21, no longer that kid who'd spent his weary days at the shanks of Guernseys. He was a grown man now full of beans.

Mama returned home to her work at the post office in Fork Union (she had given up her Rosie Riveter job in Richmond after Daddy left Fort Eustis), and Lyn got ready to ship out. He'd been in training for seven months at four bases in four states. Like the other 500 men in his battalion, he could set up, range and fire a 40-millimeter automatic weapon with perfect timing all day long, could scramble as a rifleman with the infantry, if need be, and fix his bayonet to fight hand-to-hand. They were ready to go. Maybe they'd be sent to the Pacific, where the beachhead-to-beachhead fighting was heavy; maybe to North Africa, where troops were staging for a rumored invasion of Italy. Or they might even go to England, where another invasion was taking shape. It was all part of the game of roulette the Army played. Lyn was just one of 16 million servicemen and women spinning around that wheel, his battalion antsy to get going, playing hurry up and wait.

# Camp Patrick Henry, Virginia

A week after Lyn's stroll with his bride in Savannah, the wait ended. All 500 men of the 144th AA Battalion climbed back onto Army trucks for the daylong ride up Highway 1 through South and North Carolina to their embarkation point, a staging area called Camp Patrick Henry. As before, all along the way, cars pulled over to the curb to let them pass and townspeople paused on the sidewalks to wave or salute. By then, folks were quite used to seeing long motor convoys headed north to deploy. Patrick Henry was a less imposing camp than the four Lyn had trained at so far. A flat cement-paved grid of quickly tossed together buildings hemmed in by virgin forest, the fort, which could house 35,000 transient lodgers at a time in what one soldier called "tarpaper hutments", did no combat training. It was all about getting the troops onboard their transport ships at the nearby Norfolk and Newport News docks, from which convoys all winter had been sailing for North Africa.[18]

---

[18] By the end of the war Camp Patrick Henry had processed more than a million troops, most of them headed to Western Europe.

*Camp Patrick Henry, 1944*[19]

So now they knew something. They weren't going to the Pacific theatre. They were headed back to the desert, where for months brutal fighting had been raging. For two weeks, they waited. No one was allowed to leave the base, no letters that might betray their whereabouts could be mailed. Their equipment packed and headed to port, all training exercises were canceled, except for hand-to-hand combat and a new lesson, how to abandon ship. They watched movies in the camp auditorium, played basketball and table tennis in the recreation halls, visited the libraries and the chapels, and joined in pick-up baseball on a corner lot. Mostly, they debated what they'd seen in the newsreels of the tank battles in North

---

[19] https://www.marinersmuseum.org/2015/07/inside-look-at-camp-patrick-henry/

Africa.[20] On departing Fort Bliss, the men in Lyn's battalion had hoped never to see desert again, but here they were headed back into the sand.

---

[20] See Atkinson, Rick. *An Army at Dawn*. Henry Holt: New York, 2002, a highly readable Pulitzer Prize winning history of the North African campaign.

# Atlantic Convoy

On May 9th 1943, the battalion lined up with 36th Infantry Division G.I.'s to file onboard the U.S.S. Thurston, a troop ship that had just celebrated its one-year anniversary in April. Her belly lined with bunks squeezed in so tightly that you had to turn sideways to make your way among them, the Thurston was armed with four 50-caliber guns and two 40 mm M1 anti-aircraft weapons, the very guns Lyn's team had trained on. The next day, at the mouth of the Chesapeake Bay, the Thurston joined up with eleven other troop ships (5,400 troops aboard), two tankers, and their multi-ship destroyer escort in convoy formation for a zig-zagging two-week traverse of the Atlantic Ocean to North Africa. What the men onboard didn't know was that the previous day, while they were mustering onboard their ships, the battles there had ended. A pincer-movement by an American army from the west and a mostly British army from the east (Operation Torch)[21] had trapped Nazi General Rommel's Panzer tank divisions and forced them off the continent at Tripoli on the Mediterranean

---

[21] https://www.history.navy.mil/browse-by-topic/wars-conflicts-and-operations/world-war-ii/1942/operation-torch.html.

Sea coast. This was the first battle campaign fought by American soldiers against the Nazi's and their first working in tandem with the British since World War I.

*This photo of soldiers boarding the U.S.S. Thurston troop ship was taken on May 9, 1943. Lyn may be one of the men on the dock.*[22]

---

[22]http://www.navsource.org/archives/09/22/22077.htm.

The Thurston's convoy, designated UGF-8A (United States to Gibraltor, fast ships, 8th convoy + 1), hauled troops that would form up in the desert for Operation Husky, the first step in the invasion of Europe. Probably, the men onboard had not yet been apprised of the fate awaiting them. Anyway, they had more immediate concerns. Out on the Atlantic, leaving his two Virginia's behind (the state and his bride), Lyn must have marveled at the world of water in mid-ocean, no land to be seen, the sky enormous, and the sea surprisingly calm, as the convoy ships zigzagged along their route, black smoke trailing from their diesel smokestacks.[23] For the first time, he and his shipmates were in danger of enemy attack. They crowded the deck, those not barfing with sea-sickness scanning the water for U-boats and the sky for Luftwaffe bombers.

The seamen onboard explained that no air attack would come in the open ocean. No bomber could range that far. The danger of an air attack would only arrive when they closed in on the narrow Strait of Gibraltor at the mouth of the Mediterranean Sea. The immediate danger came from German attack submarines, the dreaded U-boats (Undersea Boats) that stalked the Atlantic in "wolf packs" of up to 40 torpedo-armed submarines. U-Boats prowled the seas from Greenland to the Falkland Islands throughout the War, with free license to torpedo any ship they came across.[24] Their initial goal was to sink merchant ships conveying food, weapons and other supplies to England, but once the U.S. entered the War,

---

[23] Archival footage of the next convoy out of Hampton Roads after UGF-8A, UGF-9, underway: https://www.youtube.com/watch?v=zuRsgzTnRmQ

[24] U-Boats sank 3,000 merchant and military vessels during World War II. The Allied Navies fought back, though. Two thirds of the U-Boats were sunk in battle and the rest scuttled at the end of the War.

they began to target troop ships as well.²⁵ In response, the Allies organized multi-ship convoys escorted by a defensive shield that might include a battleship or a flotilla of destroyers with heavily armored hulls, depth charge mines and onboard artillery.²⁶ Aircraft carrier-led fleets also patrolled the high seas.

Nevertheless, U-boats wreaked havoc on shipping. Just that winter, they'd sunk 63 ships in February, several off the coast of North Carolina, and 107 ships in March, including 20 in a single convoy that had been attacked by a 40-U-boat wolf pack in the North Atlantic. In May as convoy UCF-8A sailed, there were 120 U-boats out prowling the Atlantic.²⁷ To keep their minds off all that, the troops crowded the deck for twice daily calisthenics and bull-in-the-ring hand-to-hand combat instruction. Their biggest lesson was one no war game could teach, how to cope with looming danger, when all you can do is wait.

If anyone had dreamed of a trans-Atlantic luxury tour like those in Hollywood movies, they quickly learned that troop ships offered something else altogether. The Thurston's berths, deep in the bowels of the ship, consisted of canvas and metal-frame bunks, six feet long and just two feet wide, suspended by chains and stacked one above the other as many as six high, with just a two-foot space between bunks. To climb into

---

²⁵ Interesting article in *Hampton Roads Military Museum* newsletter about Atlantic crossing convoys and their U-2 battles; includes mention of UGF-8A: https://media-cdn.dvidshub.net/pubs/pdf_49876.pdf.

²⁶ The 2020 Tom Hanks movie *Greyhound* offers a gripping depiction of U-boat assault on a WWII shipping convoy.

²⁷https://www.ibiblio.org/hyperwar/USN/rep/ASW-51/ASW-5.html.

bed required a monkey's adroitness, and once in, there was no room to sit up or roll over. Ventilation was bad and everybody smoked the free cigarettes passed out two packs to a man every day, so the air was dank and hot. To make matters worse, no troops were allowed on deck after sunset, so all they could do was lie still like sardines packed in a can while the ship rolled and the lucky men who could sleep snored. They ate beans for breakfast (competing for the most disgusting farts), supplemented with cookies and peanut butter purchased at the ship's store, and between the rhythmic heaving of the hull and the constant diesel stink, felt sick most of the voyage.[28]

As it turned out, the Thurston's convoy encountered neither U-boats nor Luftwaffe bombers on their two-week crossing.[29] The troops disembarked on wobbly sea legs a few miles east of the Algerian coastal city of Oran at a village named Mers el Kabir on May 23, at their backs the blueberry bright waters of the Mediterranean Sea and before them miles of bleak Saharan desert.

---

[28] From: https://www.4point2.org/harrison.

[29] The USS Thurston was particularly charmed in her heroic wartime service. In addition to her trans-Atlantic convoy efforts, Thurston hauled troops directly to landings at Sicily, Normandy and Iwo Jima, earning seven battle stars. Decommissioned after the war and renamed the Chickasaw in civilian service, she rests now just off the beach at Santa Rosa Island, California, grounded there in a 1962 storm.

# North Africa

*Street sign, Oran, Algeria. Photo sent home by Lyn Gentry.* [30]

Fort Bliss had been bad enough, but the North African desert was a whole other thing. Instead of checking their gear for rattlesnakes, the G.I.'s looked for camel spiders, venomous sand-colored horned vipers and thirty kinds of scorpions, four of which bore lethal stingers. To make matters worse, the notorious sirocco wind was blowing, sometimes at hurricane

---

[30] Did my father carry a camera overseas? G.I.'s weren't supposed to have them, photographs were heavily censored in the mails, yet this book includes a dozen he sent home during the war. I wonder, too, how they were developed in camp?

force, driving sandstorms northward across the Sahara Desert. During the day, anything metal was too hot to touch, and dirt rammed into mouths and ears, every crevice of clothing, weapons and the engines of vehicles. The newly arrived soldiers could hardly believe the mind-boggling store of weaponry stacked up at the docks and lining the six miles of dusty road to Oran. Acres of tanks and armored vehicles baked in the sun alongside thousands of crates of shells and all the other materiel needed for the impending invasion of Italy. A makeshift Army base outside Oran housed thousands of American soldiers, and more were arriving every day on troop convoys from Hampton Roads and New York.

Lyn's battalion was detailed to the 5th Army and the 45th AA brigade, the only American force that had seen battle against the Germans. They were put to work manning 40-millimeter anti-aircraft positions along the shore, scanning the horizon for any enemy planes that might attack from Luftwaffe bases just across the sea in nominally neutral Spain. Some platoons were sent 25 miles northeast to the port town of Arzew, where they staffed a Gun Operations Room (GOR) that coordinated Algerian coastal defenses, using the recent invention of radar and information from scout planes to order anti-aircraft batteries into action. Other platoons took up machine-gun positions 20 miles further east in the seaside town of Mostaganem. A third of the battalion's personnel got right back onboard ship to assist a convoy hauling German and Italian prisoners of war to Great Britain and the United States.[31] Though I don't know which of the three positions Lyn was detailed to, I'm pretty sure he was not on the POW mission.

---

[31] 250,000 German and Italian troops had been captured during the North African battles.

Here are the only two photos of him from North Africa, looking hale and hearty, but no longer mustachioed.

A lot has been written about North Africa by American servicemen stationed there during the War. Archival photographs feature ancient seaside towns lined with blocky Moorish buildings and winding, narrow streets teaming with G.I's bargaining with Arabs in long robes. As the Algerian-French author Albert Camus wrote (in his classic novel *The Plague*, set in a fictional city based on Oran), the drab Algerian coastal towns seemed to turn their backs on the lovely Mediterranean Sea, instead facing the bleak desert, its wind-driven dust storms, and trackless sand.

The troops were issued an Army pamphlet intended to help them understand the ways of the North African Arabs, including these pointers:
- Never smoke or spit in front of a mosque.
- Don't kill snakes or birds. Some Arabs believe the souls of departed chieftains reside in them.

- When you see grown men walking hand in hand, ignore it. They are not "queer."[32]

An online history of the famous 443rd AA Division (which fought in North Africa and all of the European campaigns) includes this jaw-dropping anecdote:

> Many Arabs were extremely poor and stealing was a way of life for them — as many G.I.s discovered to their sorrow. Stories abounded about a number of men who had been murdered for their valuables — including clothing. One 443rd G.I. awoke in the desert one morning to find that his boots had been stolen off his feet during the night. During the landings, the many ammunition trailers that had been landed at points distant from their gun-tracks were invariably found looted of everything except their ammunition. Beggars were seen everywhere. Most of the women kept the lower half of their faces veiled in public and it was noted that various tribes were identified by tattooed markings on forehead, cheek, or chin.
>
> Weekly market days were held at desert crossroads where chickens, goats, sheep, camels, fruits and vegetables exchanged hands. A battery of women could also be seen at these events, each sewing madly on a treadle-type, Singer Sewing Machine. One 443rd member, who spoke a little Arabic, was watching an auction of young girls, either sold by their fathers or captured by other tribes to be sold into slavery. Stripped to the waist, they would submit to arm and teeth inspection by potential buyers. Before he knew it the 443rd soldier had made a bid in Arabic and had bought a girl. She followed him around all day like a puppy dog but when evening came he managed to get away from her in the narrow streets — leaving her free.[33]

---

[32] Kennett, L. *The American Soldier in WWII*, p. 121.

[33] This online narrative details the exploits of an AA battalion that fought heroically in North Africa, Sicily, Italy, France and Germany:

Which leads me to the one story I heard Daddy tell about his fourteen months in Africa. We were at Lyn's Market, the rural grocery store he owned in Arvonia, Virginia, while I was in junior high school. My job was to man the front, running the cash register, gas pumps, and ice cream scoop. He worked the back as the butcher. One lazy Summer day, when he'd had a snort or two, I overhead him talking to another guy, maybe a WWII veteran himself, and crept closer to hear more. It may have been the first time I'd ever caught him saying anything at all about the War. What I heard was one of those surprising tales that can change how you see a parent.

As I recall, he stood there in a bloody apron behind the meat display case, a cigarette at the corner of his mouth, and as he spoke his voice quieted and his eyes went dreamy. He said that he liked volunteering for kitchen patrol (KP) duty in Africa, because it gave him access to dry goods, which led to an agreement with a young Arab woman, who would slip through his tent flap at night, squeeze onto his cot, and leave with a pound of sugar in a bag. It's a sordid tale, I guess, but what I remember most clearly is what Daddy said next. It seemed marvelous to him that he never saw this woman unclothed. She always wore a flowing white haik that swept the floor, and a head scarf and veil that covered all but her eyes. And then he made this surprisingly delicate gesture. He slowly waved his hand in the air, palm down, as if conjuring this sweet memory, and sighed, "Her skin was like silk."

At their battle stations, the AA battalion dined on "C rations", a bland, just palatable meat and potato hash in a can. Sometimes they'd get lucky, issued larger tins called U rations (dubbed 5-in-1 or 10-in-1 rations) with some variety that could feed five or ten G.I.'s. Or they might be issued K rations, three tightly packaged boxes of tinned meat, meat and egg or processed cheese, biscuits, crackers, dextrose tablets (switched to caramel candies later in the war), a packet of instant coffee (a new invention that became a supermarket staple after the war), a fruit bar, a chocolate bar, a bouillon cube, lemon juice crystals and sugar tablets. Supplemented with the concentrated chocolate food bar called "D" rations, and whatever eggs, produce, or the occasional chicken or mutton they could purchase in town, the servicemen got by. They learned to heat their meals on makeshift "desert stoves", half-filling empty U ration cans with sand and pouring in gasoline to boil water, make coffee, and poach eggs. A real treat was an occasional, wonderfully sweet local orange or tangerine.

Water was at a premium in the desert. Men learned to brush their teeth, shave, and dab their hairy parts with just a cupful. Not that it helped much, since the air always seemed filled with dust thrown up by passing convoys or blown in by the Saharan wind. Gun crews spent half their time greasing their weapons, and they cut parachute silk to wear as scarves pulled up over their noses like the bad guys in cowboy movies. The Mediterranean Sea lay tantalizingly close, inviting a swim, but finding a spot not hemmed in and soiled by the exhausts of the fleet at dock would not have been easy. A few lucky G.I.'s lined up to use the ancient Roman baths at a French hospital in Oran, but most of the old aqueducts along the coast had long ago dried up.

*Oran, Algeria. Photo sent home by Lyn Gentry.*

For weeks, the G.I.'s crammed into low-slung canvas tents near the Mediterranean ports waiting for the invasion to launch. No one knew which divisions would go, or where the Army would land, but clearly the war machine was gearing up. Manning an artillery battery, Lyn looked out across a sea of ships and landing vessels. The men of the 5th Army were the only

*Desert chow line. Photo sent home by Lyn Gentry.*

G.I.'s who had seen combat, having fought in Operation Torch that winter. All the other soldiers were fresh meat, as they said, antsy to get off the desert and do something that mattered. Then on July 9th, Armies from sixteen Allied nations boarded troop ships and headed out for the invasion of Sicily, an Axis occupied island in the middle of the Mediterranean Sea that looked on a map like a football kicked by the toe of Italy's boot. At 180,000 men, this was the largest invasion force in history (The D-Day invasion later in the war landed 156,000). The 5th Army led the way, as they had in North Africa, but instead of joining the invasion, Lyn's battalion was reassigned to Allied Force Headquarters and bounced first to the British 25th AA Brigade and then to the U.S. 44th AA Brigade. Still on guard duty at the Libyan ports and awaiting the return of their POW detail, which eventually arrived in mid-October, they could only watch as the mighty armada disappeared over the northern horizon.

All Summer, across the Fall and Winter, and deep into the Spring of 1944, they read of brutal battles up the boot of Italy. Sicily fell in August, but the Nazi troops there evacuated to the mainland, where they dug into fortified bunkers to repulse the Allied invaders. Fierce weeks-long sieges of the Benedictine monastery atop a hill called Cassino and at the little beach town of Anzio would make the history books, as the Allied Armies fought entrenched German divisions amidst driving rain storms on tight mountain passes. In all, the Italian campaign of World War II cost 60,000-70,000 Allied soldiers killed and 330,000 wounded.[34]

What must the men of Lyn's artillery battalion have felt, while they waited? Eagerness to join the fray? Relief at the relative safety of their positions? Worry that the war would end before they saw combat? Fear that they might soon be called on to fight? Probably all of these emotions played out during their dreary months in the desert. Here's the one message from Daddy that still survives, a Christmas post card to his favorite sister, Aunt Dorothy[35]:

---

[34] German losses: 40,000-150,000 dead and 330,000 wounded. An estimated 150,000 Italian civilians also died. Casualty figures for the Italian campaign: https://www.custermen.com/ItalyWW2/Statistics.htm.

[35] V-E cards were single sheets of thin paper on which a soldier could scribble an address and greeting. The card was then photo-copied, sent electronically in batches back to the U.S., and reprinted for mailing to the recipient. Note that this one was addressed far ahead of Christmas, on October 23.

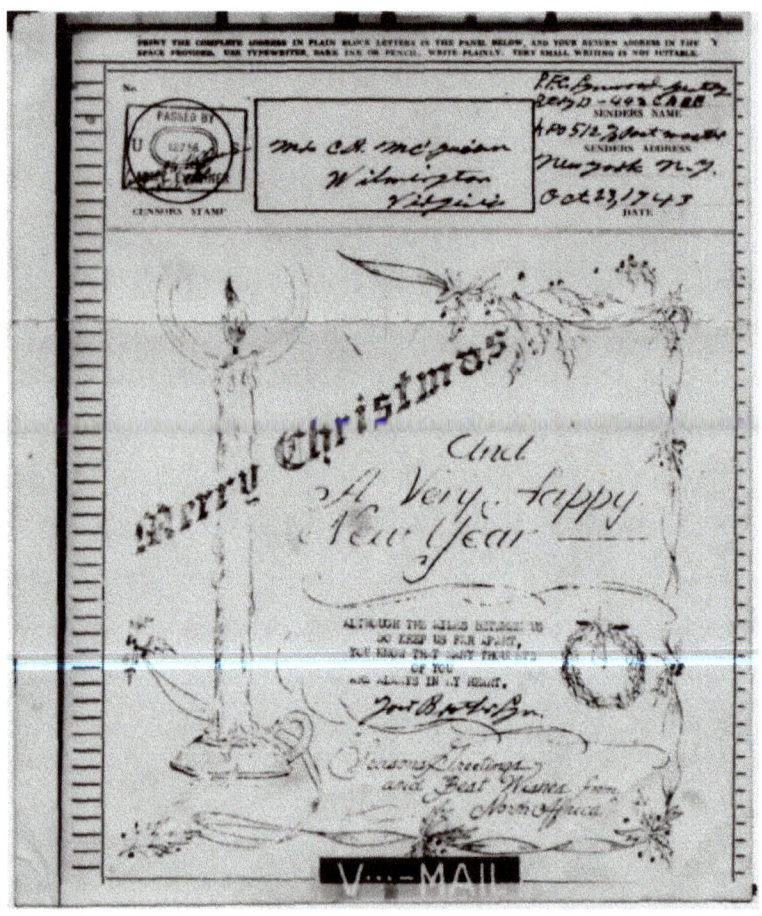

The battalion saw some movement in November, after the POW detail returned from overseas, regrouping near the Libyan capital city of Algiers, where once again they manned M1 anti-aircraft batteries overlooking the port and the busy Rehaia Airfield.[36] They rarely had

---

[36] This airfield, 15 miles outside Algiers, was headquarters of the U.S. 12th Air Force 350th Fighter Group (flying P-39 Airacobras) and the 416th Night Fighter Squadron (flying British twin-engine Beaufighters), both of which conducted intruder runs over Italy and guarded the North African coast.

anything to shoot at. The battalion's online history recounts how the crews could see German planes diving for convoys off the coast, just out of reach of their guns. There was one instance, however, when the battery in Algiers was able to zero in on enemy planes that dared to fly over the city to bomb the Guyotteville Stadium. They shouted with excitement at a direct hit on a barrage balloon, "which made a beautiful sight as it floated to earth in flames."

But that rare thrill only underscored the general boredom of their desert outpost. Whatever else the artillerymen may have been feeling, they must have felt useless as they read the tall black headline in the *Stars and Stripes* newspaper at the end of May, when U.S. General Mark Clark's Fifth Army (the one they'd been assigned to) rolled into Rome.

And then, just a few days later, D-Day, the long-rumored Allied invasion of France, kicked off. An armada of nearly 7,000 ships crowded the Normandy shore, landing a combined Allied force of 156,000 men against heavy fire from entrenched German positions. More than 9,000 Allied G.I.'s were killed or wounded in the first 24-hours of fighting,[37] but they pushed the Germans back, and the same bloody mile-by-mile advance the Allies were fighting in Italy repeated itself across Holland and France. Lyn's battalion read of a weeks-long battle at the seaside town of Caen and another dubbed Bloody Gulch. Operations Martlet, Epsom and Jupiter leveled quaint villages outside Paris. And still the artillery gunners sat in Algiers, wearily wiping sand out of their weapons, and guarding against an attack they knew would never come.

---

[37] The worst loss of life was at Omaha Beach, that landing unforgettably portrayed in the Steven Spielberg/Tom Hanks film *Saving Private Ryan*.

*Pals in North Africa. Photos sent home by Lyn Gentry.*

# Italy

Finally, in late June, their turn came. The men of the 442nd AA battalion were ordered to turn in their equipment and gather at the port of Algiers for staging to Italy. Then their orders changed, and they convoyed back to the Lion Mountain Staging Area outside Oran, where they'd first landed in North Africa. On July 22, the unit was formally deactivated and, no longer an anti-aircraft artillery battalion, unsure of exactly what they were, or what they were expected to do, on July 26 they all boarded the troop transport ship S.S. Marine Robin[38] for the three-day journey north on the Mediterranean Sea to Naples, Italy. Mama's memoir says that a troop ship on this convoy was sunk along the way. Until she got a letter from Daddy weeks later, she feared that he had been onboard.[39]

---

[38] Following her wartime service as a troop ship, the Marine Robin was carved into two boats, a barge and its tug (an excellent example of the old adage "swords into plowshares"). As the Joseph H. Thompson (and Joseph H. Thompson, Jr.), they ply the Great Lakes to this day. See photos: https://www.cbi-theater.com/marine_robin/marine_robin.html.

[39] Mediterranean troop convoys were attacked by U-boats, and ships sunk, as late as Spring 1945, but it does not appear that any ship was sunk on this convoy after all.

On July 29, after 14 months in North Africa, Lyn stepped onto the continent being called "Fortress Europe" for the first time. That initial glimpse must have been dismaying. Naples, one of the continent's most beautiful and historic cities, lay in ruins. By this time, the Italian government had switched sides, their fascist leader Benito Mussolini had been run out of the country, and there was no one left in Naples to shoot at the Americans. Instead, begging children and starving women stumbled along the docks picking over scraps of garbage. They tugged at the men's sleeves as they marched in formation off the dock. This was the aftermath of titanic battle. Not something you could ever unsee.

*Caption in Mama's Photo Album reads: "Salerno, Italy".*
*Photo sent home by Lyn Gentry.*

Lyn's artillery battalion boarded troop trucks and convoyed about thirty miles north to a training ground near the village of Piana di Caiazza, today a quaint town of stone buildings and narrow streets, overseen by a Medieval castle on a hill. The Volturno River meanders through farmland outside the town. In November, the U.S. Fifth Army had fought for two weeks to cross that river, battling on what the German's called the Volturno Line (their southernmost defensive position in the Italian campaign), but the village and the castle had been, miraculously, left untouched. After more than a year at the edge of the Sahara Desert, the rich bottomland along the river outside town, crops ripening in late summer, must have appeared like Oz, or at least a hint of home. But the artillerymen were warned not to stray; the Nazi's had strewn personnel mines all along their lines of retreat. Sappers swept the roads, flagging paths for safe passage.

On August 5th, the battalion learned (the news must have been jaw-dropping) that they would no longer use the 40-millimeter anti-aircraft guns they had been trained on in the States and had manned in North Africa. Major Gordon A. Dixon took command of what was now to be called the 99th Chemical Mortar Battalion (Motorized).[40] For the next month, Lyn's team came to grips with their stubby new weapon, the M2 4.2-inch chemical mortar, a mere pop-gun compared to the cannon they knew.[41] Nicknamed the "Four-Deuce" for its 4.2 inch wide rifled barrel of seamless nickel steel, at 360 pounds an M2 mortar[42] was too heavy to carry, so it came with its own two-wheeled hand cart. Even so, platoons were trained to break the thing down to its three pieces (tube, baseplate

---

[40] The "Motorized" designation meant that each squad was expected to haul its own weapon and ordinance, not depending on the Infantry to do so.

[41] YouTube video showing use of M1 and M2 mortars in combat movies: https://www.youtube.com/watch?v=evpJqz8SFM.

[42] History of the 4.2" Chemical Mortar: http://www.4point2.org/mortar42.htm.

and bipod) to carry by hand in case battlefield conditions made pulling a cart impossible. En route to battle, each platoon would drive its own truck and jeep, which hauled the men, their mortars, and its projectiles.

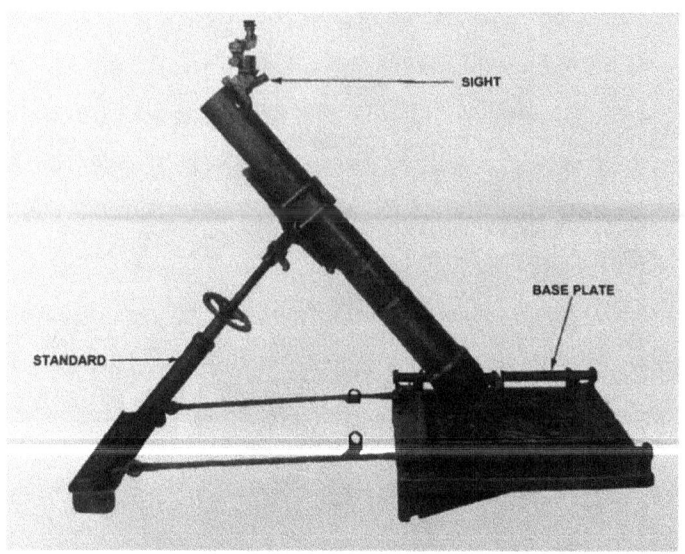

*4.2-inch M2 81-mm chemical mortar*[43]

Bemused at first, the men quickly learned to respect the fire power of their new weapon.[44] Though squat, the 4.2-inch mortar's rifled barrel could accurately fire its shells as close as 500 yards or, with less accuracy, fling them almost three miles. Each shell packed the wallop of a 105-

---

[43]https://brill.com/view/journals/vulc/9/1/article-p118_006.xml.

[44] A demonstration of the M2 mortar in action: https://www.youtube.com/watch?v=jHsICdJEp44.

millimeter howitzer.⁴⁵ Though mortars could fire explosive shells, the word "chemical" in the battalion's new name meant that their main projectile was loaded with white phosphorus, abbreviated WP and nicknamed "Willie Pete". Exploding before impact, a fine mist of white smoke instantly filled the air and began to rise, forming a smoke pillar. When shells struck side by side, they could fog out a stretch of enemy lines.

The battalion practiced coordinating their M2 mortar launches, firing in rhythm, again and again, as fast as they could, creating a wall of smoke to obscure an infantry advance. WP was scary stuff, though. The fiercely hot particles could burn through clothing, score flesh to the bone, ignite ammunition, and set everything they touched afire. Launched in conjunction with explosive shells, they fell like hellish rain on enemy foxholes. ⁴⁶

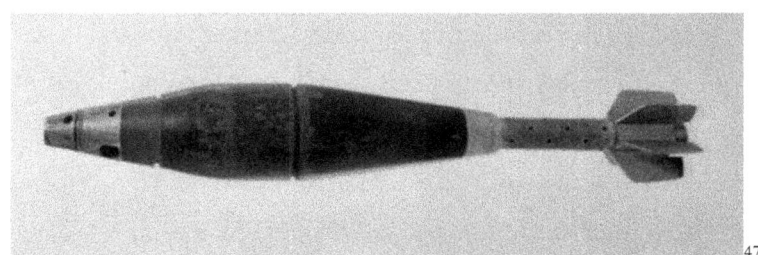

*WWII-era 81-mm mortar shell*

---

⁴⁵ For more information on the 4.2-inch mortar, see the Army's official handbook (https://bulletpicker.com/pdf/Handbook-for-the-4-2-Inch-Chemical-Mortar.pdf; and the Manual of Field Examination for 4.2 Gunners (https://collections.nlm.nih.gov/ext/dw/1308002R/PDF/1308002R.pdf).

⁴⁶ German troops called WP shells "grass-cutters", because – fired down onto foxholes -- they caused devastating fires and casualties.

⁴⁷ https://authenticpreowned.com/products/wwii-u-s-military-missile.

There was another deadly projectile available to the chemical mortar teams, kept in reserve in case the Germans dared to use it first: mustard gas. That weapon had proven so ruinous in the trench warfare of World War I (and so hard to control, as clouds of the blinding, lung-scathing, poisonous gas might drift back over one's own men) that it had been outlawed (as it still is) by international agreement. But wherever the chemical mortar battalions went, they carried shells filled with mustard gas, and others loaded with phosgene and newer poisonous chemicals. In the end, however, those shells were never used in World War II by either side.

The creation of the 99th Chemical Mortar battalion, and others like it,[48] came from lessons learned in the Italian campaign. It was clear by the Summer of 1944 that forthcoming battles would be fought at close range, as armored and infantry division actions, assaulting the Germans who were fighting behind fortified defensive lines, as they had up the boot of Italy, and were doing in France (and against the Soviet army on the Eastern front). Having highly mobile mortar units that could serve alongside the infantry, it was found, provided more accurate explosive shelling than could be provided by a distant artillery unit. Mortars also offered the welcome option of smoke screens during an open field assault. By this time, too, the mighty Luftwaffe was all but spent, so anti-aircraft platoons had little to do.

As early as 1943, General Clark had made it a policy that no infantry division would go into combat without a chemical mortar battalion

---

[48] The 100th Chemical Mortar Battalion was created at the same time, also a converted anti-aircraft battalion, the 637th AAA. The two battalions trained together in Italy.

attached.⁴⁹ Those mortarmen had shown their worth at Anzio, throwing up smoke screens to hide infantry assaults and bringing artillery-level fire

---

⁴⁹ In March 1944, Brigadier General Alden H. Waitt described the weapon in *The Infantry Journal*: "[The M2 mortar] is muzzle-loading, rifled, and fires a high-capacity shell at a high angle with the accuracy of an artillery piece at ranges from 600 to over 4000 yards. For sustained fire it can deliver five rounds per minute for an indefinite period. For short periods a rate of twenty to thirty rounds per minute can be obtained by trained crews.

The mortar consists of a barrel, a standard, and a base plate. Its equipment includes certain spare parts and accessories for its installation and maintenance, and a two-wheeled, rubber-tired, hand-drawn cart for transportation in forward areas. The barrel complete weighs ninety-one pounds. The standard weighs fifty-three pounds. The base plate is the heaviest part of the mortar, weighing 155 pounds.

The mortar shell has thin walls and large capacity and was designed especially for firing chemical agents. It weighs approximately twenty-five and one-half pounds ready to fire and holds six to eight pounds of chemical. It is prepared for firing by inserting a cartridge into its base and placing the cartridge container rings of powder sufficient to give the desired range. When the shell is loaded into the muzzle of the mortar it slides down to the bottom of the barrel where the cartridge is ignited by the striker pin. The cartridge then ignites the rings. The explosion expands the soft metal plate at the base of the shell so that the shell engages in the rifling of the barrel, thereby giving the shell a rotating flight. Inside the shell is a perforated steel vane, which causes the liquid filling to rotate with the shell and give the shell added stability in flight.

The 4.2 mortar has been used with success by mechanized troops. Installed on a mechanical mount, it has the same mobility as a tank, personnel carrier, or other track-laying vehicle. It should be especially useful to armored forces laying smoke screens to permit advance against anti-tank gunfire, or to conceal movement to attack positions."

See https//www.4point2.org/journal.htm for the full article, which includes testimonies by infantrymen on the effectiveness of this weapon in the Sicilian and Italian campaigns.

power to the front lines. General Patton, in the Summer of 1944, issued a standing order that no infantry division in his 3rd Army would be committed to combat without a chemical mortar battalion attached, and no infantry regiment would go without a mortar company. For flexibility on the battlefield, chemical mortar battalions were not made integral parts of Army divisions, but were attached to them as support where needed. They were in such high demand that often the companies of a battalion were split up and assigned to different infantry divisions. As a result, when an infantry unit was rotated out of combat, their mortar companies often stayed on the battlefield, attached to the incoming fresh infantry unit.[50]

Which is where the primary concern about switching from an artillery piece to a mortar came in. The big guns were fired from emplacements sometimes a mile or more behind any battlefield line, while chemical mortar units were imbedded with the infantry. If the mortar crews didn't lead the charge, they prepared the way with smoke screens and explosive ordnance across the field of battle, always close enough to be hit by German artillerymen determined to take out the American mortars. One private, hearing his sergeant call their role "close support" replied, "Close support, eh? You ask me, this pipe-like contraption will require a man to get too damn close!"[51]

Aiming and firing an M2 mortar looks easy, but to do it right and place a shell a thousand yards away onto a target took training and skill. Here's how one gunner described a timed test they all took:

---

[50]https://en.wikipedia.org/wiki/Chemical_mortar_battalion#cite_note-3.

[51] United States Army, "A History of the 91st Chemical Mortar Battalion" (1945). World War Regimental Histories. 85. http://digicom.bpl.lib.me.us/ww_reg_his/85.

The test started with all the mortar parts in place on a cart. They would give you an azimuth (compass bearing) and that started the clock. You'd have an assistant drive a stake about 25 yards out on that azimuth. Then you grabbed a pick and spade, and dug trenches, about five inches deep, for the base plate and the standard. You had to be right on the azimuth because the gun only had about 17 degrees of adjustment. If you cranked it that far and you weren't in line yet, you'd never have time to move the whole gun around. You hooked the base plate to the standard with two long steel connecting rods, then set the barrel in place.

They'd give you a deflection – an angle off the azimuth line – and an elevation. The elevation was the angle the barrel made with the ground. You had to finish the test with the gun assembled and dug in, and lined up on that deflection and angle.[52]

On August 23rd, after intensive mortar training with the 100th Chemical Mortar Battalion, the 99th CMB was issued forty-eight M2 mortars. They organized as four companies (A-D), further split into three four-tube platoons. Lyn was assigned to Company D. The idea was that an infantry division would go into battle supported by one or two chemical mortar companies, with twelve mortars each. When at full strength, Lyn's Company D mortar squad consisted of seven men: a noncommissioned officer squad leader, a noncommissioned officer gunner, and five junior noncommissioned officers: an assistant gunner, two ammunition handlers, and two truck drivers.[53]

Equipped with binoculars, a radio and a map, the squad leader set the mortar's battlefield location, direction and range, the two gunners took

---

[52] Eldredge. *Finding My Father's War*, p. 25.

[53] Daddy was a noncommissioned junior office, a private, throughout the War.

turns loading the mortar, and two crewmen shared in unloading projectiles from crates, adding cartridge rings to their bases, and handing them gingerly forward to the gunners. The other two crewmen acted as runners, taking turns driving the truck that held their shells (and parked far back from the battlefront), then shuttling shells from the truck to the emplacement. At least that was the initial idea. In the mountainous, muddy tracts where most of the European battles were fought, getting a jeep forward was often impossible.

Hauling an M2 mortar and its shells on foot required the whole squad. Each mortar kit came with a 2-wheeled cart that could be pulled by a jeep, but on the battlefield, marching alongside infantrymen, the squad took turns as pack mules, four at a time dragging the heavy cart through mud, up hills, across creeks, and along rocky paths. Sometimes the terrain made using the cart impossible, and in those cases, they abandoned it, three men drawing lots to see who would lift which of the three parts of the mortar (the 91 lb. tube, the 53 lb. bipod or the 155 lb. base plate) onto their backs, which was anyway preferable to the work of the other five mortarmen, who had to haul the TNT- or WP-packed shells.

The men learned that mortar shells, though heavy, were delicate. Each one had a plunger held in place by a safety pin. Before handing the shell over to the gunner to drop it into the mortar barrel, a crewman pulled that pin, so when the shell landed anything that pushed the plunger would cause it to explode. If handled too roughly, the shell's safety pin might come dislodged. To make matters worse, the ring-shaped propellant charges, packed with highly flammable nitrocellulose (extra rings were added depending on how far the crew wanted a shell to fly), were designed to detonate on impact at the bottom of the mortar tube, sending the shell whooshing out of the barrel. The men who carried those rings in their backpacks, and who fit them onto the mortars before firing, minced along with extra care.

Once in place and ready to fire, the squad needed to act with one mind to range, load and fire their weapon in a smooth, rapid and seamless flow, over and over again, sometimes launching dozens of shells in a single afternoon. The gunner slid a fat, finned projectile into the barrel, and when it hit bottom it launched with a hollow thunk, arcing high in the air and exploding on impact down-range. This all happened so fast the men didn't have time to plug their ears. And then, on either side, other mortar teams were firing, too. They hit their sleeping bags at the end of a long day with a constant ringing in their ears (Daddy came home with tinnitus that never went away, and a deafness that worsened in old age).

---

[54] Photo clipped from newsletter of the 83rd Chemical Mortar Battalion Veterans Association. https://www.4point.org/muzzleblast83/muzzleblasts-w06.pdf.

Ranging the shells was a hit-and-miss proposition at first, but before long the crews got the hang of it, and could knock out a target the size of a jeep at a thousand yards. The goal was 20 rounds a minute, one every three seconds, for the first two minutes of firing, slowing that pace to let the barrels cool, at a rate of 80 rounds an hour. It was a sort of grueling dance with live ordinance, as repetitive as an assembly line, the whole battlefield right up in your face and the infantry sprawled before you.

While in training, the 99th battalion was batted around by the 5th and 7th Armies, both of which were preparing for Operation Dragoon, the upcoming invasion of Southern France. At the end of August, as they packed up for more combat training with the infantry, they ended up assigned to both Armies. On September 6th, the 99th and 100th CMB climbed aboard trucks and convoyed north to the Fifth Army CWS Training Area, 11 miles southeast of Follonica. Along the way, they bivouacked for a night outside the town of Civitavecchia,[55] and arrived at the training ground by noon on the 7th.

Follonica today is a lovely seaside town in Tuscany, its beaches lined with tall pines.[56] I recall Daddy saying it was the prettiest place he saw overseas. But he had little time for tourism there, as the battalion immediately launched into a grueling 60-day training program, intended to get the men in shape for combat, both physically and mentally. They trained

---

[55] Today, the port town of Civitavecchia is a cruise ship stop for visitors to Rome (just 37 miles away). Featuring a tower that was partly designed by Michelangelo, the town was damaged by Allied bombing during the Italian campaign.

[56] Joseph Heller's classic war novel *Catch-22* concerns an American bomber squadron stationed in Italy at this time of the War, at an airfield on nearby Pianosa Island.

six days a week, rain or shine, under the direction of three officers who had discovered the usefulness of chemical mortars first-hand in the Sicily and Anzio battles, Lt. Col. William S. Hutchinson, Jr. (who had commanded the 83rd Chemical Battalion), his assistant 1st Lt. Alfred H. Crenshaw, and 1st Lt. George H. Young of the 2nd Chemical Battalion.

The officers laid out a hilly five-mile course along woods trails and set the men to speed marches aimed to get them in shape to walk, without double-timing, the whole route in fifty minutes. They constructed four sharpshooting ranges, three infiltration courses, and three fire and movement courses, where the men trained using live small arms ammunition. Additionally, the battalion trained on three obstacle courses, a rifle range, a bazooka range, and a hand grenade course. The idea was to get the men into top physical shape, while also learning infantry tactics.

Stepping up their training during September, the men were assigned to 7-man mortar crews and practiced the tiring ballet of loading and firing their new weapons all day long. They then trained to work as a battalion on the battlefield. All that effort paid off, too. On October 19, they held a simulated phosgene shoot, when 48 mortars placed 1709 rounds of WP (white phosphorus), HE (heavy explosive), and FS (fire support) on the deserted island of Trajaccia, off the coast of Follonica, in just two minutes.[57]

To polish off their training, at the end of October, the 99th and 100th CM battalions undertook a rigorous five-day field exercise under simulated combat conditions. The two battalions took turns acting as either a mortar group or as infantry, hauling their mortars and ordnance by hand up steep

---

[57] See https://www.4point2.org/hist-99.htm.

*Follonica, Italy. Photos sent home by Lyn Gentry.*

mountain passes in heavy rain. They camped in muddy foxholes, slept in soaked ponchos, and maybe, just for a moment, pined for the desert. Despite all this, they stayed in good spirits, eager to learn what they'd need to do in combat. All-in-all, during their two months in Follonica, they fired nearly 10,000 shells.

PULLING A 4.2-INCH MORTAR CART OVER RUGGED TERRAIN [58]

It was time. They were as ready as they would ever be. On November 8th, Lyn's mortar squad joined a truck convoy up the coast to the Tuscan port of Livorno,[59] en route to join the invasion of southern France.[60] For the next week, they bivouacked in tents at the staging area outside of town, waiting for assignment to a troop ship. The rain never let up, beating constantly on their tents and making a mud wallow of the roads, but there was time then to write a letter, drink some good Tuscan wine,

---

[58] Image: Kleber, E.K. *The Chemical Warfare Service*, p. 187.

[59] Livorno means "Leghorn" in Italian; that's where the leghorn chicken was first bred; on Allied maps, the town went by the name Leghorn.

[60] The 100th CMB stayed behind and fought with distinction in the rugged Dolomites of northern Italy right up until the end of the War. More info: https://www.4point2.org/hist_100.htm.

and get their muddy uniforms cleaned. To do that, you stripped down in one tent, tossing your filthy clothes into a bin, then strode naked through the rain to the next tent's communal shower faucets, and finally to a third tent where an orderly passed you a neatly folded clean uniform. Sometimes, if you were lucky, new boots that might actually fit awaited you, too.

*Private Lyn Gentry, Italy, 1944.*

# On to France

They began to load the U.S. Merchant ship JAMES MOORE[61] on November 16th and were all ready to ship out three days later. After laying at anchor with all onboard another day, the ship began its run west along the Italian coast on the way to southern France, escorted by a single French corvette.[62] It was a choppy voyage, with mistral winds churning heavy waves. The ship's captain sought shelter at the harbor of Madealene on the northwest coast of Sardinia for a day, but the ship got stuck on a sandbar in port, troops piling on deck to watch a tugboat eventually pull her into a safe anchorage. The next day, with slightly better weather, they set out again, sailing through the Straits of Bonifacia and hewing close to the west coast of Corsica.

---

[61] Originally a cargo ship, the James Moore had been converted to a troop ship that could carry 350 men sleeping in tight quarters on 3-tier bunk beds. At war's end, the James Moore was reconverted to carry grain on the Great Lakes and was eventually scrapped in the 1970s. See this comprehensive article about WWII troop ships: https://ww2ships.com/acrobat/us-os-001-f-r00.pdf.

[62] Corvette's were the smallest WWII Allied vessels designated "warships". Corvettes sank 42 German submarines in WWII. More info: https://militaryhistorynow.com/2020/08/17/the-flower-class-corvette-nine-facts-about-the-tiny-warship-that-played-a-huge-role-in-ww2/.

SS James Moore

The ship dropped anchor at the outer harbor of Marseilles with a seasick load of G.I.'s on Thanksgiving Day, and rocked on its moorings for another day. Finally, the following evening, the ship docked and began to unload its weary troops. Lyn's squad climbed in a truck and joined a nightlong convoy uphill through the winding French Riviera mountains to their staging area 18 miles north of the city.

No telling what they saw of France's second-largest city in the middle of the night. The old port of Marseilles was in ruins, dynamited by the Germans before they were chased out of town during the Operation Dragoon invasion in August (11,000 Nazi troops surrendered).[64] The whole harbor and its inlets were mined, but minesweepers had cleared the shipping lanes and wharfs by the middle of September, when the rest of the

---

[63] Image: https://www.armed-guard.com/lsip02.html.

[64] See this day-by-day chronicle of Operation Dragoon: https://history.army.mil/brochures/sfrance/sfrance.htm.

invasion force began to arrive. In any event, the port was fully functional by the time the 99th CMB rolled ashore.

At the staging area outside Marseilles, the men of the 99th CMB fought wind and rain to set up their pup tents, spending the next few days collecting equipment and supplies from the ship. Then the battalion boarded World War I-era 40 x 8 boxcars (called that because they were originally configured to haul 40 soldiers or 8 horses) for a 3-day train ride up the Rhone River valley en route to the Seventh Army. All the while, the mistral that had tossed their ship on the passage from Italy howled, pushing temperatures far below freezing.

Luckily, that first night, they parked at a French barracks in the town of Valence, where they slept on straw out of the cold. By the next night, they'd made it all the way to Dijon. They unpacked their pup tents to bivouac on wet grass, but then ran across men from the 12th Chemical Maintenance Depot, who allowed the whole battalion to crash in their warehouses overnight. Back on the tracks the next morning, they reached their destination, 7th Army Headquarters, by nightfall. They bivouacked in pup tents that night in a pasture 10 miles west of Epinal, in the foothills of the Vosges Mountains.

The battalion stayed there only a couple days, sent on by 7th Army orders to Sarrebourg,[65] where they were attached to the XV Corps. Companies A, C and D bivouacked near the hamlet of Hommarting. Company B camped nearby in Veustholtz. This was the Alsace-Lorraine region in the far eastern corner of France, a bounteous farmland of cheese, beef, pork, wine and beer, though the conquering Germans had long ago

---

[65] Not to be confused with the larger town Saarburg, Germany, which is fifty miles north near Luxembourg.

raided the best of the foodstuffs there. The western border of Alsace-Lorraine rippled like a shaggy carpet, the forest-covered Vosges mountains running north to south in a long line dotted with picturesque hamlets occupied since Medieval times and nestled in winding river valleys, those mountains receding to rich eastern farmlands and vineyards that stretched flat as a hand out to the German border, marked by the deep, wide and fast-moving Rhine River. For the next four months, shuttling from one skirmish to another, Lyn's battalion would fight in a zig-zag pattern all over this region, engaged in house to house conflict amidst a bewildering array of towns that covered all of Alsace-Lorraine from the western slopes of the Vosges Mountains to the Rhine River border with Germany.

*Pvt. Lyn Gentry (center) with pals*

# The Accident

As a young man, yours truly introduced wine to our family's Southern Baptist Thanksgiving dinners that had hitherto featured only iced tea. And one year, I even splurged on a bottle of after dinner cognac. Daddy sniffed the stuff and quaffed it (we didn't have snifters, but a jelly jar worked in a pinch), and told one of his few war stories, about how it seemed every farmhouse in France had a cellar carved out of limestone, lined with dusty bottles of wine or champagne and, yes, a burning golden liquid that tasted something like the elixir in his glass. Cognac was his favorite, high octane, the true treasure. He was a man of few words, so we kids had learned to extend them by the look in his eyes. In telling this tale, they danced.

Apart from that giddy anecdote, Daddy never had a kind word for France. He said it was dirty, that people pooped right in the street. He'd seen a statue that he couldn't believe, of naked little boys peeing into a fountain. Or maybe that was Italy and he'd mixed the countries up. Or maybe, more likely, I'm remembering it wrong. Whatever the case, France was where an accident happened that almost got him court-martialed. It dogged him the rest of his life. I've heard versions of it from both parents, but can't recall who said what now. Anyway, this is what I remember.

After church on Sundays (his one day off work), when the weather was good, Daddy would head off on a stroll down the tree-lined railroad tracks near our house, and I'd tag along. Sometimes we'd take fishing poles to a pond back in the woods; more often we'd turn off on an overgrown path to an abandoned old farmhouse, where he'd sit on the porch with its Black caretaker Elmo,[66] and shoot the shit. Most of the time, I'd go off and kick up grasshoppers in the field, but sometimes I'd sit and listen. That's where I heard him tell this story, later confirmed and embellished by Mama, about what happened. His eyes looked haunted when he said these four words: "I killed a Frenchman."

It happened at night, very likely a rainy night, on a winding country road in wine country. Daddy was driving a truck back to camp from town. (I never heard what town or why he was out there at night.) A local farmer appeared in his headlights, but Daddy saw him too late. It's easy to imagine the scene: Broken body in the rain, dented fender, panicked Army PFC radioing in to his command post, military police, handcuffs, a makeshift stockade set up in a barn. Whatever else happened during the War, I believe this was the turning point, the event that came to represent in Daddy's tortured heart all that followed.

He'd killed a man.

In the end, he was not court martialed, though Mama said it was a close call. Probably, the War saved him; he was needed at the front. His punishment instead was losing the Private First Class stripe he'd earned in

---

[66] After watching the movie *Mudbound* (recommended), I wondered if Elmo, too, might have been a WWII vet, and if those Sunday afternoon chats may have been a kind of therapy for both men? Sure wish I'd listened more closely.

Africa, knocked back to buck private (his $54/monthly pay reduced by $4)⁶⁷, which is the rank he held at war's end.⁶⁸ All of which leads me to consider the road not taken. What if he had been convicted? He might never have seen combat, all the rest of the story I'm about to tell, and all the other events that haunted him. But "what if" is a fool's game always, isn't it? They let him out of the stockade, sent him back to his unit, and very shortly they all went into battle.⁶⁹

---

⁶⁷ In today's dollars, $54 amounts to $911; his pay cut amounted, in today's dollars, to $68/month.

⁶⁸ Here's how I figured this out. Note that on the card sent to Aunt Dorothy from Africa his rank is "pfc". PFC's wear a single stripe on their sleeves, while buck privates wear no insignia. In the photos of Daddy in uniform, taken during his leave home in 1945, his sleeves are bare, and his discharge papers state his rank as private.

⁶⁹ A caveat here. Daddy's unit served in France for four months, most of it under battlefield conditions. I'm placing the incident here, ahead of battle, but really it could have happened at any time while he was in France.

# December 1944

"It's almost impossible to learn about combat before you're in it."

U.S. Army Medic, WWII Brendan Phibbs[70]

It's hard to say how much G.I.s on the ground in Europe ever knew about their commanders' grand plans. From camp to camp and battle to battle, they followed orders and coped, often without any granular notion of why they were being sent where, beyond the key compass point: to our east is the Rhine River. On the other side of that river is Germany. Your job is to push the Nazi's back across that river and take Hitler out. Go. At some point, however, their commanding officers must have pulled them together for a show-and-tell, offering a version of what the generals of the European theatre saw from their map rooms in commandeered chateaus behind the lines. At the end of November 1944, this is what they saw:

---

[70] Phibbs, Brendan. *Our War for the World: A Memoir of Life and Death on the Front Lines in WWII.* The Lyons Press: Guilford, CT, 2002.

*From a biweekly series of wartime maps.*[71]

The light gray area on the left of the map represents a million Allied troops who were pressing forward in Belgium and France. Having already taken Paris, they moved deep into champagne country, where the web of trenches from World War I were being put to new use. Another quarter million had landed in southern France and fought their way north around the neutral border with Switzerland and the imposing mountains of the French Alps, turning east now towards the German border on ground between the Rhone and Rhine rivers that had been enriched by the blood

---

[71]   https://en.wikisource.org/wiki/Atlas_of_the_World_Battle_Fronts_in Semimonthly_Phases_to_August_15_1945

of warriors going back to the beginning of European history. American, Canadian, English and even a French Army formed a ragged line north to south down the length of France, taking one fortified village at a time, losing it again to German counter-attack, and pushing on when they could. The 99th CMB was assigned to the Seventh Army, forming the southern stretch of this line in Alsace-Lorraine, a region that had been traded between France and Germany for generations, where as many people spoke German as French, and where butter-and-wine French cuisine deliciously melded with the sausage and beer appetites of Germany. To the south, skirmishes continued in the Italian mountains, where the 99th CMB had trained, and to the east the relentless Russian army (light gray on the right side of the map) was pressing hard into Poland and Yugoslavia, with Berlin in its sites.

On the map, Hitler looked finished, his vaunted Third Reich sandwiched between mighty armies (the black ink on the map shows contested territory; the white ink shows Germany and its captured regions; the dark gray ink shows neutral countries). The CO standing with his pointer at the front of the command tent would have turned away from his map at that point to remind the G.I.'s in attendance that -- just as they'd been apprised way back in boot camp -- now that the Nazi rat was cornered it would get up on its hind legs. Now, they warned, the fight begins.

Lyn had been in uniform for two years. His battalion had missed the fighting in North Africa, the invasions of Sicily and Italy, D-Day in Normandy, and the Operation Dragoon assault on southern France. But on December 1, they found themselves camped just 10 miles away from the German 19th Army and 40 miles west of Germany, on the front lines.

This book is about grunts on the ground, not generals in confiscated castles, but I think it's important to explain a controversial decision by Allied Supreme Commander General Eisenhower that governed everything Lyn's unit did as part of the Seventh Army (Sixth Army Group) in Alsace-Lorraine. Before the 99th CMB arrived at the front, the Seventh Army had already fought its way up the Rhone River from Marseilles, climbed into the Vosges Mountains and pushed the Nazi forces all the way back to the German border, capturing the Rhine River city of Strasbourg. Sixth Army Group Commander General Devers saw a daring opportunity to cross the river and stab deep into Germany, but Eisenhower would not let him do it. The Supreme Commander preferred to line up all of his forces on the border with Germany – the English Army in Belgium, General Patton's Third Army in northeastern France, and the Seventh Army and French First Army in Alsace, in order to wipe out all Nazi resistance in France before crossing into Germany itself.

That decision has been debated ever since it was made. Critics, including General Devers himself, have argued that a swift strike into Germany would have scrambled Hitler's hard-pressed forces, so they would not have been able to organize for the Ardennes Mountains assault now known as the Battle of the Bulge and the concurrent Vosges Mountains battle known as Operation Nordwind. The European war might have ended months sooner, preventing the deaths of as many as 40,000 Allied soldiers. We'll never know, of course, because instead of pressing east across the Rhine River into Germany, General Devers was ordered to split the Seventh Army, one half stretched along the Alsatian plain, charged with holding ground already gained on the French side of the Rhine, while the other half fought north through the Vosges Mountains to bolster the flank of General Patton's 3rd Army, which was bogged down in mud in the Lorraine region.[72] So, in the winter of 1944, skirmish by skirmish in farming hamlets along the German border, the final chapters of the European war were written.

In line with those grand plans, the 99th CMB camped in the northern foothills of the Vosges Mountains. Company B joined the 44th Infantry Division, setting up a line of mortars outside the Nazi-occupied town of Durstel; Company C set up outside nearby Greis, with the 79th Infantry Division. Lyn's Company D joined the 100th Infantry Division, bivouacking at forest edge and siting towards Zittersheim and Moderfeld, mirror-image villages of pale pink stucco buildings clustered around old stone churches in a narrow valley nestled amongst ridges lined with hardwoods, bare of leaves at the end of autumn. Though they looked idyllic, these towns were in fact fortified defensive positions protected by German infantry and Panzer tanks with their dreaded 75-millimeter cannons, along with mortars and heavy artillery emplaced on mountain redouts. Machine guns bristled from high windows, snipers hid in bell towers, and the roads into town had been heavily mined. Villagers cowered in basements and wine cellars, praying.

Historian Alex Kershaw wrote that no army in history had successfully battled across the Vosges Mountains, adding:

> It was hard to imagine a worse place to fight. All the advantages lay with the Germans. Jagged ridgelines overlooked deep valleys pitted with peat bogs, leading to windswept moors. Giant granite tors dotted thick conifer forests at higher altitudes. The tightly bunched firs towered more than seventy

---

[72] This long-debated decision, and the angry arguments among the generals leading up to it, is neatly detailed in Clarke and Smith's *Riviera to the Rhine*, pp. 437-445.

feet in places, and without a compass, men could get lost after crawling beneath the lower branches for only a hundred yards. There were few roads and those that could hold tanks were narrow and twisted sharply as they snaked up and down, from one village of half-timbered homes to another. And the Vosges were famous for their atrocious weather, even for northern Europe, with cold fronts settling over them for weeks on end.[73]

To make things more difficult, the mountains ran along the Maginot Line, a seven-mile deep swath of steel-reinforced concrete bunkers, linked underground tunnels, barbed wire and trenches that stretched the length of the Franco-German border. This defensive wall had been constructed in the decades since World War I to keep the Germans back (Hitler's armies, in the 1940 invasion of France, had simply bypassed it, entering France through Belgium to the north), but now, ironically, the Maginot Line was in the hands of Nazi forces, who were using its fortifications to defend against Allied assault on their border.[74]

The 44th and 100th Infantry Divisions had first fought in mid-October along the Maginot Line. The 2nd battalion of the 44th had battled with particular distinction in a skirmish near Sarrebourg, fending off a picked Panzer division from shallow foxholes, even while outmanned and outgunned. Seventh Army commander General Patch later credited the 2nd battalion with saving the whole Seventh Army that day. By the time the 99th

---

[73] Kershaw, Alex. *Against All Odds*. Dutton Caliber, New York, NY, 2022.

[74] Today near the town of Siersthal, you can tour an underground network of bunkers, tunnels and pillboxes that made up part of the Maginot Line fortifications at Le Fort Invicible du Simserhof: https://www.cc-paysdebitche.fr/tourisme-culture-sport/le-simserhof/

CMB joined their infantry units, those troops were battle-hardened and well-schooled in the tactics of mountain town combat.[75]

The men of the 99th CMB quickly learned that working with infantry divisions meant being fully imbedded and constantly on the move. In town-to-town and door-to-door skirmishes, they had to pull their mortar wagons or sometimes haul their ordnance on their backs, scurrying to any barrier they could find and setting up at close range in the midst of battle. Was sixty days of intensive training enough to turn artillerymen into battlefield mortarmen? Was five days of simulated combat enough to prepare them for the real thing? The men of the 99th CMB were about to find out. For the first time, in early December, these artillery platoons turned mortarmen found themselves smack in the middle of it all.[76]

Getting shot at was terrible. As was passing through a village blown to rubble and corpses by your own ordnance. Daddy said this, more than anything else, is what haunted his dreams, memories of the ragged aftermath of bloody battle, town after town and day after day. The online history of the 99th CMB states: "This was the first time this unit had seen combat and it received its 'baptism of fire' without any preliminary combat indoctrination. Morale was very high and the men showed determined initiative from the beginning."

On their second day in combat, after helping the 79th Infantry Division clear Gries, Company C took the battalion's first prisoner of war,

---

[75] From a history of the 44th Infantry Division: https://www.sonsoflibertymuseum.org/44th-infantry-division-ww2.cfm.

[76] A YouTube video showing use of M1 and M2 mortars in combat movies: https://www.youtube.com/watch?v=evpJqzZ8SFM.

when a Frenchwoman in the town of Wahlenheim pointed to a barn where a Nazi soldier was hiding. They moved on to take Kurtzehausen, then were ordered to switch infantry divisions, joining up with platoons of the 44th to take seven more villages (Mackwiller, Butten, Diemeringen, Enchenberg, Maierhof, Petit Rederching, and Siersthal). Company B, meanwhile, fought along with the 44th Infantry Division through seven towns (Lohr, Adamswiller, Waldambach, Butten, Hammerkapt, Rohrbach, and Petit Rederching) in just two days. And Lyn's Company D fought with the 100th Infantry Division through Hangwiller, Moderfeld, Zittersheim, and following a particularly fierce door-to-door skirmish, Wingen,[77] before they, too, were reassigned to the 44th Infantry Division on December 4th, pressing on through six other towns (Volksburg, Rosteig, Montbronn, Enchenberg, Heilingbronn and Siersthal) over the next week. No one mortar platoon did all of this, of course, as they were each working with different battalions of their assigned infantry divisions, snatching a fitful nap in one captured town before moving on to fight their way through the next.

Chemical mortar battalions were in short supply on the battlefield, and regimental commanders juggled them as needed. Here's how that juggle looked in just three days of early December. On their fourth day of battle, Lyn's Company D was attached to the 114th Regiment of the 44th Infantry Division (XV Corps), while the rest of the 99th CMB went with the 324th Regiment. The next day, Company B and one platoon of Company C were reassigned to the 324th Regiment, and the rest of Company C to the 71st

---

[77] A 100th Infantry Division rifleman's account of his experiences in combat on the Franco-German border, including mention of battles in Rosteig and Enchenberg: https://warfarehistorynetwork.com/article/love-company-in-the-voghes-mountains.

Regiment. Company C was reassigned four days later to the 114th Regiment with Company D and for the first time came up against German tanks.

Most of the battles fought in the Vosges Mountain campaign have gone unrecorded, but this one, in the sleepy town of Enchenberg, may serve as an example of how infantry and mortar squads collaborated in those skirmishes.[78] The 114th Regiment was a New Jersey National Guard unit that had first engaged in battle in mid-October as part of the invasion of Southern France and had fought its way across the Vosges Mountains since then. Their assignment on December 7 was to capture Enchenberg, a picture postcard town of 80 or 90 stone and stucco buildings clustered around a steepled church situated on a high ridge in a gap between two heavily forested hills. It was assumed that the village was occupied by Nazi troops, but no one knew for sure, and no reconnaissance of the town had yet been made.

Wintry rain fell as three infantry companies (each composed of about 100 men) advanced on the town. The Germans waited until they were just 50 yards away before opening fire with machine guns. At that, the G.I.'s fell back and the mortars of Lyn's Company D already set up at wood's edge dropped explosive shells on the area, silencing the machine guns. But then the Germans started launching mortars, too, their shells exploding in the midst of troops seeking shelter on the nearby hills. As German shells fell around them, Company D switched to WP, laying down a smoke screen to shield a second advance across the field. Louder than all this firing came a

---

[78] This account drawn from a narrative of the battle at: https://warfarehistorynetwork.com/article/delaying-action-at-enchenberg.

terrific explosion. The Germans had blown up a road bridge over a railway in the center of town.

*Google Map of 44th Infantry Division route
December 4-12, 1944*

*Eglise-Saint-Pierre, Enchenberg, France*

Infantrymen reached an old stone house at the edge of the village just as a Panzer tank turned the corner and began to pepper them with its 75-mm cannon and machine gun. Inside the house, radioman called out coordinates for the mortars and for a 105-mm howitzer parked on a nearby hill. Their shelling chased the tank away. As the G.I.'s went door-to-door clearing houses, around noon they saw a line of Germans setting up near the railroad bridge and again called in mortar and artillery fire, killing or wounding some of the Nazi squad and forcing the rest to seek cover. That's when the Panzer tank reappeared, blasting armor-piercing shells into the houses where the Americans had taken shelter. Again, they called mortar fire down, forcing the tank to retire.

Meanwhile, Infantry Companies B and C took cover in woods on either side of town, while the men of Company A hunkered down in houses overnight. Before midnight, the Panzer tank returned and began to methodically knock down walls with shots from its 75-mm cannon, eventually retreating again before dawn. Somehow, the weary G.I.'s got through the rainy night.

The next day's plan was for Company A to hold tight while the other two infantry companies pressed in on both sides of town, clearing houses door to door. But German mortar and small arms fire pinned down the troops in the woods. Guessing that observers in the steeple of the town church were directing fire, an M36 tank destroyer pulled up at wood's edge and fired three 90-mm rounds at the steeple. After the third shot, the German mortars stopped firing.

About this time, a second German tank appeared, lining up alongside the first, but both tanks retreated under bazooka fire. As the three infantry companies worked through the town, they again faced mortars and machine guns, and scurried from house to house, gradually gaining ground and pushing the Germans back, until the Panzer tank once again charged their way, firing its cannon point blank at any wall still standing in the center of town. A bazooka shot failed to slow it, then another took out a side wheel of its tread, but the tank came on, soon rejoined by the other tank. All the while, German infantry in the shadow of the tanks skirmished with the Americans, both sides wielding machine guns and grenades.

Before sunset, the German tanks withdrew again, leaving most of the old stone houses of Enchenberg in ruins. While American artillery pounded the north end of town where the Germans were holed up, the G.I.s tried to catch a nap where they could. They hoped for American tanks to roll in by morning, but two of the eight Shermans headed their way were damaged by mines, and the others halted, afraid to risk the roads. Instead, a no fire line was drawn through the middle of town, and at dawn 96 guns, including 105-mm and 155-mm artillery and Company D's M2 mortars began to pound the German-held north end of town. When the firing stopped, the Germans were gone, and the G.I.'s marched through the ravaged village unmolested, except for occasional sniper fire. The men of the 114th Infantry Regiment had shown they could take on German Panzer tanks without tank support of their own.

The men of the 99th CMB had shown that they could fight, too. This may have been Lyn's first big battle, one that reduced a pretty village to rubble. Exhausted and with ears ringing, he trudged through what was left of Enchenberg. He saw the wounded stacked on jeeps, the dead piled onto trucks, and the townspeople returning from their hideouts in the woods to crushed and burned homes. And he learned that a German mortar shell had landed on a squad from his battalion's Company C during the fighting, wounding five men and killing another.[79] The first casualties of the 99th CMB, they had been in combat less than a week.

The rest of the 99th CMB was earning its combat pay, too. The next day, the 1st platoon of Company B was taking shelter in a captured house in Petit Rederching, when a Nazi sergeant armed with a light machine gun

---

[79] The 99th CMB online history is confusing on this point, stating: "The first casualties suffered by the battalion occurred in Company C on 10 December in the town of Enchenberg, France. A short burst from one of the mortars injured five men and killed one man." This could mean that an American shell exploded in its tube, rather than that a German shell struck the mortar crew. In an anecdotal history of another mortarman, *Finding my Father's War*, author Walter J. Eldredge describes the risk of faulty mortars, especially in cold weather, and mentions a casualty-causing barrel burst suffered by the 99th CMB (p. 197).

challenged them. Two mortarmen -- PFC Charles F. Wyatt and Pvt. Francis M. Girard -- took the man's gun and made him a prisoner of war.

*99th CMB mortar crew firing on Enchenberg, December 1944.*[80]

Two days after clearing Enchenberg, amidst a similar engagement near the town of Lambach, is when it must have happened. I don't recall how I learned this. Maybe Daddy said something about it, but I think it was Mama who allowed, under my insistent questioning, that she had heard this story. Anyway, at some point during the fighting, Daddy dropped a shell into the barrel of his mortar, leaned back as it launched, turned to his buddy for the next projectile, and saw his friend's head taken cleanly off by an enemy shell. The shell exploded nearby, wounding six other squad mates, but somehow left him unharmed. You can only imagine it. Men fallen and bleeding, crying for help. Lyn standing there stunned. He'd known this man since Texas, had shared an antiaircraft battery with him in Oran, a pup tent

---

[80] From https://warfarehistorynetwork.com/article/bloody-fight-for-hill-351/

in Italy, and barnyard haystacks in France. His best friend, lying headless in the foxhole, collar gushing.[81]

The 99th CMB had only been in combat for ten days, in tit for tat battles along the Maginot Line, and had already seen twelve men injured and two killed. But as the battalion's online history says: "Throughout the drive with the 44th Infantry Division, the 4.2" mortars were very effective in screening the infantry advances, reducing enemy strong points, and forcing the enemy to withdraw. Several direct hits were made on enemy self-propelled weapons (tanks and tank destroyers), troop concentrations, machine-gun nests, and motor transports." Very quickly, it seems, the 99th CMB was learning to fight.

On December 14, with no rest, the 99th CMB was reassigned again, relieving the exhausted and undermanned 83rd Chemical Mortar Battalion[82] that had served with the 36th Infantry Division throughout the Italian campaign, including the battle of Anzio, and during the invasion of Southern France. Lyn's battalion had been at Fort Pickett, VA with the 36th and they had convoyed together to North Africa, so this was a reunion of

---

[81] I placed this personal anecdote here, because it matches the online battalion history that says "On 12 December, casualties were sustained in Company D when seven men were injured by fragments from an enemy shell in the vicinity of Lambach. One man died as a result of this shelling." This is the only death in Company D specifically noted in that history. The document mentions other fatalities later in the war, but doesn't say in which company they happened. Note also that mortar squads were typically made up of seven men. Seven crewmen were casualties in the December 12 shelling, but Daddy was not injured at any time in the war, which may argue for his friend's death on a later occasion.

[82] History of the 83rd CMB: https://www.4point2.org/hist-83.htm.

sorts, though the battle-hardened infantry veterans must have wondered whether these newly blooded artillerymen turned mortarmen could fight.

The 36th Infantry Division was then headquartered in Ribeauville, a mid-sized city of cobbled streets and ribbed stucco dwellings overlooked by a hilltop Medieval castle that had so far avoided bombardment. Less than 15 miles west of the Rhine River border with Germany, Ribeauville sat along the northern boundary of a bulge in the German line known as the Colmar Pocket, named for Colmar, the French city at the center of that bulge. Just as the 44th Infantry Division had been tasked with pushing the Nazi Army back from the Maginot Line into Germany, the 36th was joining in the effort to "burst the boil", closing the Colmar Pocket and forcing the enemy out of France east across the Rhine River.[83] The 36th Infantry Division had been fighting a grinding back and forth battle for the past week against a larger German force, their positions frequently overrun, and the platoons of the 83rd CMB sometimes cut off from their infantry attachments, firing until they ran out of mortar shells and then, taking heavy casualties, skirmishing with rifles. By the time the 99th CMB relieved them, the 83rd CMB was down to a single platoon.

The 99th CMB assembled near Ribeauville and bivouacked with their respective regiments in the farming towns of Adamswiller, Bettwiller and Ottwiller on December 15 only to turn around the next day and truck convoy a few miles north to St. Croix Aux Mines, where they were

---

[83] By war's end the 36th Infantry Division had spent 400 days in combat, accepted the surrender of Field Marshal Hermann Goering, won seven campaign streamers for its colors, taken part in two assault landings and 14 of its members had won the Medal of Honor. The Division had the ninth highest casualty rate of any Army Division in World War II. See: https://www.texasmilitaryforcesmuseum.org/texas.htm.

reassigned from the 36th to the 3rd Infantry Division, under the operational control of the French Army. By this time, they must have felt like pieces in a game of speed checkers played by mad generals.

If not mad, the generals were at least confused, because the previous day the German Army in the north had launched an enormous surprise attack through the heavily forested Ardennes Mountains, an all-out assault intended to drive a wedge between the American and British Armies and probe to the Atlantic Coast, in order to take the important port city of Antwerp. Hitler thought that a swift and overwhelming strike might succeed, and even if it didn't, it would force so many casualties among the Allies that the British and American publics might cry for an armistice. General Eisenhower quickly responded, ordering the Seventh Army then fighting along the Maginot Line to move north to cover ground left undefended by the 3rd Army's move against the Germans in what was to be known as the Battle of the Bulge. The American forces at the Colmar Pocket were stripped of men and supplies for the northern battle, and the divisions that were left were ordered to hold in place or retreat if necessary.

*Colmar Pocket (dotted line). Note that the Rhine River marks the Franco-German border. Note also that the western edge of the Pocket is deep in the heavily forested and fortified Vosges Mountains.*

These were orders that angered General Patch of the Seventh Army. He did what he was told, though, forging a piecemeal fighting force along a 126-mile front of the Rhine River with only six infantry divisions (roughly 20 miles of front to each division, six miles per regiment, and two per battalion), and joining the French Army in confronting the Colmar Pocket area on three sides. His ace in the hole was the 3rd Infantry Division, renowned from World War I as "The Rock on the Marne." Battle tested

---

[84] From Prefer, Nathan N. *Eisenhower's Thorn on the Rhine: The Battles for the Colmar Pocket,* 1944-45. Philadelphia: CaseMate, 2015.

and well-schooled in assault tactics, they were ready. Though the 36th Infantry Division had taken a defensive posture along the Maginot Line, the 3rd, supported by the 99th CMB, began to probe the northern edge of the Colmar Pocket.

The weather, however, was miserable. In mid-December, temperatures that had reached the 50s earlier in the month began to plummet. It was cold, wet and overcast, so planes couldn't fly. In the flooded Alsatian plains where they bivouacked, vehicles of all kinds found it impossible to navigate the narrow roads. Tanks, especially, bogged down in the mud and didn't dare risk the rickety bridges in the area.

The colder it got, the more equipment began to malfunction.[85] Parts of the M2 mortars, such as elevating screws and springs, broke far too often, leaving units short of weapons. The screws seemed to break because units were firing their mortars beyond maximum range, putting the weapons into unusual firing positions or using additional propellant. The springs, however, were criticized as being poorly made. If the spring did not work right, it placed added stress on all the other components of the weapon. Spare parts quickly became scarce. A third issue to arise was with the ammunition. Shells were exploding either in the tubes or just after exiting upon firing. Suspect ammunition was impounded, but the problem continued to occur. Faulty fuses were the culprit, and artillery fuses were eventually substituted.[86] Fortunately, on the 19th, a detachment of the 12th

---

[85] From Kleber, Brooks E. *The Chemical Warfare Service: Chemicals in Combat.* Washington, DC: Center of Military History, United States Army, 1990.

[86] Whenever possible, mortar squads would pack sand bags around their mortars, reaching over to drop a shell into the barrel, then ducking behind the sand bags in case of a barrel explosion. But in the fast-paced battles they were fighting, it was rarely possible to set up their positions so carefully.

Chemical Maintenance Company, consisting of one officer and 16 enlisted men, came to the rescue, their sole purpose being to maintain and repair mortar equipment.

By Christmas week, Lyn's Company D, now attached to the 15th Regiment of the 3rd Infantry Division [87] (one platoon was detailed to support the 254th Infantry Regiment), was back on the line, fighting mud and Germans in fruitless efforts to cross the L'Ill River, a tributary of the Rhine. If the Americans were to succeed in closing the Colmar Pocket, they needed to cross that river, but the Germans held fortified positions at every bridge and crossroads town and the river had flooded across the flat lowlands, freezing into a sheen at night, then melting again into a muddy morass by mid-morning. In the trenches, fox holes filled with water, clothing was sodden, and soldiers crowded into barns to briefly dry themselves as best they could before heading out to fight again.

When tourists visit in the summertime these days, Alsace-Lorraine and its rustic villages, vineyards and farms can seem like the prettiest region of France. But at the end of 1944, its ruined towns, blasted roads, and

---

[87] From website: https://www.historylink.org/File/10353: During World War II the 15th Infantry Regiment, 3rd Infantry Division, earned 16 Medals of Honor. This was more than most divisions. 15th Infantry Regiment Medal of Honor recipients include Audie Murphy, the war's most decorated and famous infantryman. The 3rd Infantry Division won 38 Medals of Honor, the most of any division. This heroism came at a high cost; the division lost 4,922 killed in action and 636 who died of wounds. These were the highest losses of any division in the war. The 15th Regiment suffered 1,633 killed in action and 419 missing in action.

*99th CMB pal with M2 mortar, bazooka and snow.
Photo sent home by Lyn Gentry.*

miserable winter weather were anything but pretty.[88] A disgusted General Patton even wrote in a letter to the American Secretary of War: "I hope that in the final settlement of the war, you insist that the Germans retain Lorraine, because I can imagine no greater burden than to be the owner of

---

[88] Fascinating then-and-now photo album of Alsatian towns: https://tracesofevil.com/search/label/Sigolseheim.

this nasty country where it rains every day and where the whole wealth of the people consists of assorted manure piles."[89]

And then it began to snow. Relentlessly. All day and night for days at a time, piling chest-high drifts often whipped by arctic winds that howled down the Rhine River valley and through the mountain passes. Tanks, trucks, jeeps, and infantry platoons stood out like targets on the whitened fields. Expecting winter weather, the Nazi's had painted their equipment white and showed up to fight in waxed white coveralls. It took weeks for the American army to supply the same, so GI's raided farmhouses for sheets and towels they could use to fashion makeshift camouflage ponchos. Digging foxholes at the edge of fields gone blank as a sheet of paper, they broke their shovels on the hard-frozen mud.

From the Ardennes forests at the Battle of the Bulge to the Alsation farmlands where Lyn's battalion labored, the warriors at the German border shivered all night and awoke each day to the coldest, snowiest winter in at least 50 years. Historian Stephen Ambrose, in his WWII history *Citizen Soldiers*, asks, "How cold was it? So cold that if a man didn't do his business in a hurry he risked a frostbitten penis. So cold the oil in the engines froze. Weapons froze. Men pissed on them to get them working again, a good temporary solution but one that played hell with the weapon."[90]

And then the Germans attacked in force. No longer content to hold defensive lines, the Nazi's launched Operation Nordwind, an all-out offensive against the northern edge of the Colmar Pocket, intent on

---

[89] General George S. Patton, Jr., letter to Secretary of War Henry L. Stimson (quoted in Prefer, *Thorn*, p. 7).

[90] Ambrose, Stephen. *Citizen Soldiers*, p. 372.

pushing upriver 30 miles to the city of Strasbourg, in order to make that victory a Christmas gift to Hitler. In a pincer movement, a second German army pushed towards Strasbourg from the north, squeezing the Seventh Army in between. On Christmas Eve, the Germans launched frontal attacks from the towns of Bennwihr and Sigolsheim, which they had heavily fortified as the last villages that stood between the American Army and the city of Colmar. Just 50 miles from their Ruhr Valley war munitions factories, the Germans seemed to have an endless supply of tanks, artillery and other weapons. The American mortar battalions, meanwhile, were hoarding their own shells, waiting for new shipments that never seemed to arrive.

*Denson, the name on the jeep, must be Bronze Star awardee Pfc. Joseph E. Denson. Guessing he's the guy on the right. Unidentified officer (?) on left. Photos sent home by Lyn Gentry.*

When the attack came, the 3rd Infantry Division, stretched out across the Colmar Pocket border, was caught by surprise, but they regrouped, counter-attacked, and on Christmas Day pushed the Germans back from Bennwihr. It helped that for once the skies had cleared, allowing Allied bombing sorties against German positions.[91] Still, fighting raged into the evening. Despite their capture of Bennwihr, the Allied Supreme Command instructed the Seventh Army to pull back if they had to and let the Germans have Strasbourg, since no troops or weaponry could be spared from the Battle of the Bulge up north.

The 3rd Infantry Division did not pull back. While northern units fought at Bennwihr, the 15th Regiment, supported by Lyn's Company D of the 99th CMB, spent Christmas Eve and Christmas Day locked in savage battle on a knoll called Hill 351, overlooking the town of Sigolsheim.[92] Most of the area was flat farmland, but one bare, irregular mass of rock dominated the Alsatian plain. German artillery dug in atop the hill could observe and fire mercilessly down on anything that moved in and around Sigolsheim, so taking the town required first taking that hill.[93]

Lt. Col. Hallet Edson, commander of the 15th Infantry Regiment, realized what his men were up against: "This miniature Cassino was defended by 200 crack SS troops under orders to hold their positions to the last. These men were stalwart, fanatical, and determined. With six machine guns covering the slopes and abundant artillery and mortar fire,

---

[91] On Christmas Eve 1944, Allied air power flew 7,000 sorties in Europe, dropping 10,000 tons of bombs and destroying 116 enemy planes.

[92] Lyn's Company D was the only company in the 99th CMB to take part in this battle.

[93] Today, a necropolis and war cemetery crowns this hill. It commands astounding 360 degree views, and it's easy to see the tactical superiority any spotter units would have up there. https://www.visit.alsace/en/230104703-national-necropolis.

they constituted an extremely formidable force."[94] At dawn on Christmas Eve morning, Edson's regiment, supported by Lyn's Company D of the 99th CMB, began their harried climb. Company B commander First Lt. George Mohr recalled, "We were going one foot at a time. Each time we moved, a rain of accurately directed artillery and mortar fire fell on the men. As we pressed forward, we encountered fire from half a dozen machine guns, which had excellent fields of fire; they dominated our approach to the crest. All of a sudden, one of my officers got a wound in the chest. I knew that we were in for one hell of a battle. The fighting was so bad that Company B was eventually forced to dig in."

In war movies, soldiers do impossibly heroic things, climbing out of their foxholes, daring their comrades to join them, and racing forward amidst heavy shelling to take out machine gun nests and direct fire on the enemy. Sometimes, in real wars that happens, too. At Hill 351, First Battalion's commander Lt. Col. Keith Ware did all that and more. All morning his men had struggled on the hill, taking heavy casualties and eventually hunkering down in any shallow ditch they could find on the bare slopes. Ware got up and climbed the hill alone, deliberately drawing fire on himself to help pinpoint enemy positions. Then he went from foxhole to foxhole, urging his men forward. Only ten men dared to go with him, as he led a tank in a daring assault on the six machine gun nests atop the hill. Boldly advancing under heavy fire, he took out two machine guns by himself, and shot tracers at two others, so the tank gun could find and silence them. Behind him, five of his ten companions fell wounded. Ware

---

[94] From a quite thorough narrative of the Sigolsheim battle: https://warfarehistorynetwork.com/article/bloody-fight-for-hill-351/

was shot in the hand but battled on. For this action, he was later awarded the Medal of Honor.[95]

Having taken out four of the six machine gun nests atop Hill 351, the Americans felt the tide of battle turn. Right behind the charging infantrymen, Lyn's mortar platoon crested the hill and set up as dozens of Germans, their positions shattered, frantically ran downslope towards Sigolsheim, their numbers decimated by American mortar fire. As the smoke cleared to a rosy sunset over the Vosges Mountains that Christmas Day, what was left of the 15th Regiment stood atop what they later dubbed "Christmas Hill" (the Germans called it "Bloody Hill"), staring down at their next battlefront, Sigolsheim. It had been a life and death struggle; no one had slept in three days; dead and injured men were being cleared from the slopes for the rest of the night. Company B had taken the brunt of it, losing all of its officers and over forty percent of the company. One G.I. recalled, "They made that attack on that hill with over a hundred men and it took them two days to get the Germans off that hill. They said it was devastation, they lost a lot of tanks, it was a real war they had on that hill. The Germans had everything, tanks, artillery, and everything on that hill."[96]

Set up in the hilltop ruins of the German fortifications, the mortar platoons fired WP shells all the next day, spreading walls of smoke around Sigolsheim to screen the infantry's advance into the village, as a jeep drove up and down the back of the hill bringing shells from a supply truck parked in the woods. The mortarmen then switched to explosive shells and cut

---

[95] Eventually promoted to the rank of major general, Ware was killed in action 25 years later, while commanding the 1st Infantry Division in Vietnam.

[96] https://americanveteranscenter.org/2013/10/interview-charles-oneil-tec-5-3rd-infantry-division-15th-regiment-company-2nd-platoon/

loose against enemy strongpoints, tanks, and troop concentrations. In the village, the men of the 15th Regiment fought a relentless two-day house-to-house skirmish, annihilating the crack Nazi Zeiher Battle Group and SS Battle Group Braun and taking over 500 prisoners. In the process, Sigolsheim was reduced to smoking rubble. One of the most decorated units in the U.S. Army, having already fought heroically in North Africa, up the boot of Italy, and in the invasion of Southern France, the 3rd Division ranked the battle of Sigolsheim among its toughest battles.

Though undermanned and outgunned, by the last week of the year, the 3rd Infantry Division had pushed back the German assault all along the northern edge of the Colmar Pocket. By New Year's Day 1945, supplemented by the 254th Regiment of the 63rd Infantry Division on their left flank, the 3rd Division held a 15-mile wide line blocking the way to Strasbourg. The men of the 99th Chemical Mortar Battalion had in the month of December shown the G.I.'s the power of their weapon. One infantry officer wrote: "We have yet to see an enemy position that was tenable when we fired on it with WP and HE from this mortar. They can reach into almost perfectly defiladed positions, and their effect is devastating. A great advantage lies also in the fact that the projectile is silent, and still you have the effect of a 105-mm shell burst with greater rapidity of fire. I think it is the finest weapon in existence."[97]

The online narrative of the 99th CMB summarizes their first month of combat this way: "The weather remained at the freezing point throughout the period and the ground was heavily blanketed with snow. The mountainous terrain made operations very difficult. During the month

---

[97] Eldredge. *Finding My Father's War*, p. 90.

of December 1944, the battalion fired 15,178 rounds of HE and 10,890 rounds of WP shells. Twenty-three enlisted men were battle casualties during the month, with two of these resulting in deaths."

*The ruins of Sigolsheim after the Christmas Day battle.*

---

[98] https://warfarehistorynetwork.com/article/bloody-fight-for-hill-351/

# January 1945

Effective New Year's Eve, the 99th Chemical Mortar Battalion reorganized to make up for the men lost in December. Company A was dissolved, its members filling gaps in the other companies, and Lyn's Company D was redesignated Company A, which would be its title for the remainder of the war. The battalion's forward command post remained at Ribeauville, with the 3rd Infantry Division, and the rear post where casualties were managed was at Ste Croix Aux Mines. The battalion's companies found themselves stretched out with the regiments of the 3rd Infantry Division on a zig-zag line from just west of Orbey to Sigolsheim, Ostheim and Guemar at the northern edge of the Colmar Pocket, in a stalemate for control of the Kaysersberg Valley. After a month of combat, the 99th CMB may have hoped for a New Year's respite, but that was not to be. Having failed in their effort to capture Strasbourg as a Christmas present for Hitler, the German 19th Army pressed their Operation Nordwind offensive, designed to split the American 7th Army aligned along the northern edge of the Colmar Pocket and the French 1st Army on the southern edge and retake Strasbourg. Failing that, Hitler hoped to at least draw Allied forces away from the Battle of the Bulge in the Ardennes mountains, so his troops there could push on to the sea. And failing that, he wanted as many Allied soldiers killed as possible. Hitler told his generals:

> This attack has a very clear objective, namely the destruction of the enemy forces. There is not a matter of prestige here. It is a matter of destroying and exterminating the enemy forces wherever we find them. The question of liberating all of Alsace at this time is not involved, either. That would be very nice; the impression on the German people would be immeasurable, the impression on the world decisive.... But that is not important. It is more important, as I said before, to destroy his manpower.[99]

The 99th CMB had fought their way through Christmas, missing the hot turkey dinners shared by battalions to the rear, and now all planned festivities for the new year were ordered postponed, too. Unlike the mid-December Nazi offensive in the Ardennes, Allied generals saw the renewed Operation Nordwind coming. Intelligence about an impending attack was passed down to frontline soldiers, though the size and power of the enemy's force remained a mystery. As snow fell heavily in 3-below temperatures, the troops dug trenches, prepared foxholes, planned and cleared fields of fire, planted minefields, unrolled barbed wire and laid obstacles along roadways.

In relentless snow and sleet during the first few days of the year, Lyn's battalion, which had not yet fought defensively, faced a determined Nazi assault. On January 5th, the German army launched an attack along

---

[99] From a *History Network* narrative, *Operation Nordwind: The "Other" Battle of the Bulge.* https://warfarehistorynetwork.com/article/operation-nordwind-the-other-battle-of-the-bulge.

*Google map of the defensive line occupied by 3rd Infantry Battalion
Along the northern edge of the Colmar Pocket
New Year's Day 1945*

the Moder River at the quaint hillside town Wingen,[100] engaging raw recruits from the 70th Infantry Division in a 36-hour house to house battle,[101] before establishing a road block across Route 419, an important east-west supply road. Next day the Germans captured a Maginot Line fort near Philipsbourg. At the same time, small enemy forces estimated as about a battalion crossed the Rhine near Herrlisheim.

The enemy commenced to shell Saaralbe, Saar-Union, and Strasbourg with 280-mm cannons, some of the biggest guns used in the War. In a heavy snowstorm that day, the 3rd platoon of 99th CMB Company

---

[100] Private Frank H. Lowry's (276th Infantry) harrowing account of the 36-hour house-to-house battle for Wingen:
https://www.trailblazersww2.org/units_276_wingen.htm.

[101] *History Network* account of Wingen battle:
https://www.historynet.com/battle-of-wingen-sur-moder/

C received such heavy shelling that it was forced to evacuate its mortar positions at Orbey and move back. By the 7th the Germans had extended their Rhine bridgehead to a 6-mile front from Drusenheim to Gambsheim. On the 8th, a bitterly cold day with the thermometer down to zero and in the face of a sharp wind, the Seventh Army ordered a retreat to the Maginot Line, forcing the 99th CMB's forward command post back to Lapoutrie. The enemy followed, reaching the Maginot Line north of the Haguenau Forest by nightfall.

Next day, in a blizzard blowing blinding snow down the mountain passes, the Germans launched a methodical series of attacks against the Maginot Line, with their main effort towards the adjacent towns of Rittershofen and Hatten, both of which were defended by entrenched American forces. With the ground frozen, German tank battalions rolled right across open fields, overrunning Allied positions in the towns of Boofsheim, Rossfeld, Herbsheim, and Obenheim, where 300 American soldiers were captured. By this time, most of the quaint villages along the Maginot Line lay in ruins, the rubble of stone buildings providing some cover for anti-tank squads, but there were just too many tanks to hold off, the German Panzers rolling forward in mighty phalanxes and emerging from the snow like devouring monsters. Town by town, the Nazi's continued to expand their Rhine bridgehead, recapturing villages they'd lost in December. The tank battle for the fortified town of Hatten – the last great armored face-off of World War II - was especially bitter.[102]

---

[102] Summary of the battle for Hatten: https://military-historian.squarespace.com/blog/2020/10/20/an-alsatian-hellscape-the-battle-of-hatten-rittershoffen.

Part of the American origin story still taught in schools is the saga of Valley Forge, that bitter winter of 1777-78 when General George Washington's revolutionary army was encamped in deep snow. The privations of that winter are taught as a crucible where our patriotic mettle was sorely tested. But historians recognize that the ordeal suffered by Allied troops in Alsace-Lorraine in the winter of 1944-45 may have been worse. In the first place, Washington's men lived in cabins they built themselves, kept warm by fireplaces. They could rest assured, as the blizzards howled, that no surprise attack was coming. The G.I.'s crouching in foxholes along the snowy Maginot Line had no such assurance. Distinguished historian Max Hastings explains:

> Every soldier spent far more time digging than shooting. It required the labor of many weary hours to contrive a hole deep enough to shelter a man effectively from shellfire. Within days of creating such a refuge, he was required to move on and repeat the process. Soldiers performed every natural function in the open; ate clumsy alfresco picnics of nourishing but monotonous food; lived in filthy and often damp clothing that went unwashed for weeks, even months; and were subject to the arbitrary authority of those appointed to lead them. It became a luxury to enjoy the occasional opportunity to occupy quarters in a ruined building, a few days billeted in a farmhouse....[103]

In the field, huddled next to a shivering buddy in a hole in the ground, soldiers were lucky to get four hours of fitful sleep a night. While the American Army rarely fought after sundown, the Germans seemed to relish nighttime skirmishes, launching artillery fusillades at odd hours, sending out snipers and infiltration squads, and using the cover of darkness to reposition their lines. Campfires were ruled out, because they helped Nazi

---

[103] Hastings, Max. *Armageddon*, p. 140-1.

gunners zero in their guns. The artillery shellings, when they came, were horrible. Howitzer shells exploded in the treetops, bringing down heavy limbs and shrapnel together; mortar shells dropped straight down almost silently, their detonations leaving gaping craters. One soldier recalled that undergoing an artillery barrage — in Alsace such shellings might occur periodically all night — "was something inhuman and terribly frightening.... It is like the finger of God." [104]

For the men of the chemical mortar battalions, undergoing artillery and mortar shelling must have played additional mental tricks, since they were experiencing first-hand the same bombardment they had been inflicting with their own weapons. There was no getting around it, the winter of 1944-45, was a snowy hell.

On January 10th, a little late one might say, crates of "spook suits" began to arrive on the front. The American soldiers donned the white capes and painted their helmets, mortars, and other equipment white. They also received shipments of snow-ready "shoe pacs" that helped keep their feet a little warmer and drier.[105] The spook suits reduced casualties from enemy fire, and the shoe pacs reduced casualties from frostbite and trench foot. All the while, as the 99th CMB's website says: "The companies of the 99th Chemical Mortar Battalion supported the [3rd Infantry] division and attached units, firing on enemy patrols, known enemy positions, tanks, machine-gun nests, motor transports, and enemy troop concentrations."

---

[104] Kennet, Lee. *The American Soldier in WWII*, p. 136.

[105] Shoe pacs were calf-length moccasins with a leather top and rubber foot, far and away the best boots for cold, wet weather.

Town to town skirmishes make sense on paper, but the fighting in Alsace in January must have approached chaos. The whole idea of a border between the warring armies in villages lining winding mountain passes in blizzardlike conditions would have been farcical. An American veteran of the 79th Infantry Division said, "It was a weird battle. One time you were surrounded, the next you weren't. Often we took refuge in houses where the Germans were upstairs. We could hear them and vice versa. If they didn't make a move, we left, and if we didn't make a move, they left."

Historian Stephen Ambrose noted, "Flamethrowers were used to set houses afire. Adding to the horror, the civilian population had hidden when the battle began and now the women, children, and old folks huddled in the cellars. There was no electricity. The pipes had frozen so there was no water. The soldiers on both sides did what they could to feed and care for the civilians.... There was hand-to-hand fighting with knives, room-to-room fighting with pistols, rifles and bazookas." A German soldier recalled a skirmish in the town of Rittershoffen, where "almost all of the buildings, including the church, were in ruins. Many of the houses were on fire and lit up the scene at night. The dead lay about the streets, among them many civilians. The cows bellowed in their stalls, unattended, the cadavers of animals stank and infected the air."[106]

---

[106] https://warfarehistorynetwork.com/article/operation-nordwind-the-other-battle-of-the-bulge/

*3 pals in winter whites. Photo sent home by Lyn Gentry.*

To make matters more chaotic, the Germans were sometimes using American tanks which they had captured, just as the Americans commandeered German tanks. In rear areas, tanks roamed about the snowy woods, seeking mischief or a way back to their respective lines. German tanks mistook American infantry for Germans; some Americans followed German tanks, thinking they were American. The forward zone, several miles deep, was full of parties of both sides who from time to time unexpectedly encountered one another. Everybody was wearing white in the midst of blowing snow, so telling squads apart was always a challenge.[107]

An anecdote by an American soldier describes the situation in the freezing foxholes during Operation Nordwind:

> My squad spent that night in a snow covered clearing, in deep fox holes which had been dug by a supporting artillery unit; they and their guns had been pulled back to a safer location. The bottoms of the holes had been lined with empty brass 105-mm shell casings, which offered a little protection from the icy bottom of the holes, that is until the ice broke, the shell casings sank and our shoes were 4 inches into the icy water. It was really hard to get to sleep standing up, with cold, wet feet I could hardly feel the blisters on my heels.[108]

In the midst of that bedlam, on the 11th, German forces cut off and captured 1,000 American soldiers in the town of Obenheim, then switched directions and captured 700 more near Benfeld. By January 17th, the Germans were pushing deep into the Parc Nord de Vosges towns that Lyn's battalion had fought so hard for in December, but in snowy mountains and running out of men and supplies, their attack stalled.[109] On

---

[107] https://tradocfcoeccafcoepfwprod.blob.core.usgovcloudapi.net/fires-bulletin-archive/1945/MAR_1945/MAR_1945_FULL_EDITION.pdf

[108] Anecdote from: https://battleofthebulge.org/2013/12/24/the-maginot-line-and-operation-nordwind/

[109] By the 18th, what there was of a battle line ran along the Rhine River from Strasbourg to opposite Erstein, then west of a point just south of Erstein as follows: Ill River – Selestat (held by Americans) – Benweiner (American) – Sigolshein (American) – Turckheim (held by Germans) – Munster (German) – St. Amarin (American) – Thann (American) – Mulhouse (American) and Kembs (American).

the 19th, for the first time all month, the skies briefly cleared, and American fighter-bombers roared in, strafing enemy forces and dropping more than 100 tons of bombs. But the Germans regrouped, breaking through 36th Infantry Division defenses at Rohrwiller[110] and Drusenheim and capturing whole battalions of Americans.[111]

Coming under Germany artillery fire as they set up their mortars outside one town, the men of Lyn's mortar platoon dived for cover when shrapnel struck their stack of shells, causing white phosphorus to leak out and catch fire, burning into their ammunition. Decades later, Corporal Byron D. Lemmon recounted, "I grabbed a shovel and threw it in the creek but some more rounds came in and a piece of shrapnel hit me in the chest. I had a cut in my uniform but with everything I was wearing [to stay warm] it saved my life."[112] For this action, which may have saved his platoon, Lemmon was awarded a Bronze Star, along with the first of his two Purple Hearts.[113]

Hard-pressed by the Wehrmacht assault, Sixth Army Group Commander General Devers decided that the best defense was to go on

---

[110] Medical doctor Brendan Phibbs' memoir *Our War for the World* includes his day-by-day journal kept during these battles, providing a riveting account of the January skirmishes and their cost in casualties.

[111] A novel that appears to capture the maddening chaos of this sort of battle is Donn Pearce's *Nobody Comes Back* (New York: Doherty Associates, 2005). Recommended.

[112] https://www.dvidshub.net/news/129438/world-war-ii-veteran-visits-former-unit

[113] Lemmon died in April 2022, almost 99 years old, just as I was beginning to work on this story. Probably the last of the 99th CMB to go, and another of my "if only's."

the offensive. He ordered a pincer attack on the Colmar Pocket from north and south aimed at the major Rhine River bridge crossing at the old fortress town of Neuf-Brisach. While the Germans concentrated their Operation Nordwind on recapturing Strasbourg, the Allies could hit them by surprise at the Colmar Pocket. If the Americans from the north and the French from the south could capture the Neuf-Brisach bridge, they'd have the Germans in Colmar trapped, their supply lines cut, and they could use that bridge to flow into Germany themselves. Against Operation Nordwind, Devers gave his plan the optimistic title Operation Cheerful.

Devers knew that the brutal winter weather would make assault more difficult, with temperatures hovering around freezing and up to three feet of snow on the ground in mid-January. He knew that the defending Germans would be able to shelter their troops indoors, in as yet unbattered hamlets inside the Colmar Pocket, and that they could channel tanks and infantry reinforcements across the Neuf-Brisach bridge, providing a crucial advantage. He might have delayed the assault, but Allied weather forecasters predicted a thaw coming in early February that would turn the area into a muddy quagmire, complicating vehicular movement. The Allied armies were already suffering in frigid foxholes, fighting almost daily skirmishes to hold the Germans back from Strasbourg and the Maginot Line. It was time.

The men of the 3rd Infantry Division hated fighting defensive battles. They were an assault division to a man and ached to take it to the Germans. On the 19th, they got their chance when field orders arrived outlining their role in attacking Neuf-Brisach from the north. Lyn's Company A was again assigned to the 15th Infantry Regiment (with one platoon supporting the 254th), Company B went with the 30th Infantry Regiment, and Company C went with the 7th. In preparation, they practiced quick drills in speed marches, field firing, river crossing, night tactics, and how to use German weapons, all amidst heavy snow. The division's 10th Engineer Combat Battalion assembled bridging equipment.

The assault began on January 22nd, one year to the day since the 3rd Infantry Division had landed at Anzio. Eager to be on the offensive, the division's commanders called their part of the battle plan Operation Grand Slam. Part of their strategy was hoping that the Germans would be tricked into thinking they were attacking Colmar itself, and line up to defend that city, while the real target was the bridgehead a few miles east of Colmar at Neuf-Brisach. The German army was not the only thing that blocked their way. Four icy rivers and the broad Colmar Canal lay across their path, with every bridge and crossroad mined, manned and fortified against assault.

To make matters worse, the attack began in a blinding snowstorm. The troops donned heavy clothing under their spook suits, then piled on four bandoliers of bullets each, along with three fragmentation grenades, a day's rations, a blanket, a shelter half and their M-1 rifles. The mortar squads hauled all that while dragging their mortar carts through the snow. As they moved out, the blizzard was blowing hard at a daytime high temperature of 14 degrees. They trudged along in whiteout conditions, burdened by heavy gear, through deep snow drifts. Except for patches of pine woods here and there, the terrain was generally open, snow blowing across flat, frozen farmland that stretched for miles. Good tank country.

On the first night, the 30th Regiment crossed the Fecht River from their base at Guemar and moved towards a bridge over the Ill River outside the German-occupied hamlet of Houssen. In the darkness, engineers reinforced the rickety bridge as best they could, but the first tank across collapsed it. As day broke, the Germans warmly sequestered inside the town's walled Chateau de Schoppenwihr awoke and began to fire on the bridge, forcing the G.I.'s back. The Americans dug in and were able to hold their position on the river bank, while their engineers worked all day, under persistent enemy fire, to reinforce the bridge. At last they were able to fight their way across, but German tanks and swarming infantry pushed them back again to the north side of the river. Forward observers and radio operators for Company C of the 99th CMB had to swim the icy waters to escape capture.

Setting up along the river bank, the mortarmen bombarded the German tanks and the villa itself, then switched to WP ordnance, laying a smoke corridor that enabled the G.I.'s to cross the bridge and a mile of open field into the village, which – after heavy house to house fighting – they took by day's end, allowing the 30th Regiment to claim the sprawling chateau as a base for tanks and other heavy vehicles. The 99th CMB website says, "The mission was very successful and the vehicles loaded with personnel and supplies moved in without being fired upon." At another bridgehead of the Ill River, the 99th CMB's Company B maintained a smoke screen all daylong so engineers could install bridges strong enough for tanks to cross.

Lyn's Company A, meanwhile, got caught up in a brutal two-day battle for the Maison Rouge bridge across the Ill River and the nearby town of Holtzwihr. Three regiments of the 3rd Infantry Division, including the 15th Regiment to which Company A was attached, emerged from a forest west of the river on the night of the 23rd, planning to build a tank-capable bridge across the river, which was about 90-feet wide at that point. When they discovered a wooden bridge already there, they tried to run a tank across it, but the bridge shook so badly that they abandoned that plan and set out to build their own bridge after all. The engineers went to work, knowing it would take all the next day to get the job done.

While they hammered, infantrymen were sent across the old bridge on foot, their tanks and armored vehicles idling on the west bank. Their mission was to somehow hold off armored assault by the Germans, without having any armored support themselves, long enough for a tank-capable bridge to be built across the Ill. They couldn't see far in the blinding snow, but they could hear the grinding gears of approaching Panzer tanks somewhere across the frozen fields. There were problems at the bridge; important sections failed to arrive, and as the snowy day darkened towards nightfall, the first tank to attempt a crossing collapsed the whole thing.

Then all hell broke loose. On the open plain with their backs to the icy river, the G.I.'s were sitting ducks for what hit them, a whole German Panzer tank battalion and the infantry regiments at their flank. They fell back against a wall of 57-mm and machine gun fire, many swimming the river to escape sure death or capture. American tanks on the west bank fired back at the attackers, forcing a stalemate for one night. Eventually, both sides ceased firing. Their soaked clothing freezing to hard shells on their backs, the American G.I.'s shivered in the dark, huddling beneath idling tanks for the warmth given off by their engines. Others fought the cold by frantically scratching shallow foxholes in the frozen earth, knowing what awaited them come sunrise.

As expected, morning brought a fierce artillery exchange across the river. Somehow, a few units of the 3rd Infantry were able to get across to the east bank on a bridge north of the fighting, pressing forward through snowy woods to the town of Riedwihr, a key north-south crossroads that was heavily fortified. At woods edge, the Germans unleashed a cyclone of tank and artillery fire from the village, forcing the men to take what shelter they could find in holes dug by the exploding shells, where they spent another frigid night, under periodic artillery barrage.

By this point in the war, the Germans had devised a deadly system for repelling infantry assault. Sequestered within the stone walls of Alsatian towns, they first launched fusillades from 88-millimeter cannons that were either free-standing or tank-borne. As one G.I. recalled, "the aggressive resonance of the German 88's ejaculatory sounds was unique, not duplicated, to my knowledge, by any other artillery piece in World War II. It had the hoarseness of a deadly cough, the baritone echo of thunder, and you could hear it coming. Whump. Whump. Whump."[114]

Typically, the 88s fired in a straight, ladder-like pattern that could cover a 500-yard depth into Allied lines, shells exploding every thirty yards along each step of that ladder, then repeating the pattern backwards, all in less than a minute. The trick, when the firing began, was to dive into a foxhole or fall flat in between the ladder's rungs, so the shells stepped over you.

When the 88's stopped, mortar shells began to fall, raining down with a brief whistling sound as if dropped from the heavens, and making craters in a scattershot pattern as far back as a half mile behind the front line. G.I.'s learned that it was no safer to hide than it was to advance, so they moved forward dauntlessly across open farm fields that surrounded fortress towns, just as machine guns began to fire, scanning left to right as they came. The chemical mortar companies, ears still ringing from the

---

[114] Kotlowitz, Robert. *Before their Time: A memoir*. Knofpf: New York, 1997.

preliminary shelling, went to work laying smoke screens on the fields to hide the infantry's advance.

If the G.I.'s could make it into the town, then they faced door-to-door fighting with rifles, machine guns, bazookas, grenades, and sometimes fists and knives. If they were forced back, then the German Panzer tanks came forward onto the field, flanked by Wehrmacht soldiers, to finish the job. At Riedwihr, the 15th Infantry Regiment and Lyn's Company A of the 99th CMB faced just such a relentless barrage.

*99th CMB crewman cold-testing a 4.2 mortar near Ostheim, France, January 23, 1945*

By this time, the Army engineers had laid enough bridge so a few tanks could fight their way across, and three joined the embattled regiment

---

[115] https://www.reddit.com/r/WW2info/comments/11nbqfr/a_us_m2_42inch_chemical_mortar_is_demonstrated_at/

outside Riedwihr. As a dim morning light sifted through fog and light snow, Company A's four mortar platoons, set up in artillery craters, began to exchange mortar barrages with the Germans. When the American tanks edged forward across the open field, two were hit immediately by artillery shells and burst into flames. Crewmen bailed out of their hatches with clothes on fire, rolling in the snow amidst a hail of gunfire. The third tank pressed on, but its gun was broken. Its commander bravely turned it broadside to the German assault so his crew could pile the smoldering crewmen from the other tanks onboard, retreating back to the woods under heavy fire. German foot soldiers came charging across the field, and the fighting turned into man-to-man duels in the woods that lasted the rest of the day. Though not mentioned in the battalion's online narrative, Lyn and the other mortarmen must have abandoned their mortars and joined in the fray. At nightfall, the Germans pulled back to the shelter of town, while the surviving Americans hunkered down in the icy woods, collecting their dead and wounded.

Before dawn, the battered 15th Infantry unit and Lyn's Company A were ordered to abandon their position at Riedwihr and trudge south through a forest to Holtzwihr, a neighboring village which sat between the Ill River and the Colmar Canal. If the Americans were to reach their objective at the Rhine River bridgehead of Neuf-Brisach, those streams had to be crossed, and in order for bridges to be laid across them, Holtzwihr had to fall. Even if the regiment had been at full strength, this was a suicide mission. Like Riedwihr, Holtzwihr was heavily fortified, and without tanks and armored vehicles, the Americans gathered at the edge of a sparse pine woods were sitting ducks, stuck waiting for the inevitable German assault once the sun came up.

*Google map of Holtzwihr; note Ill River to left (west), forest at top center, Riedwihr at top right and Colmar Canal running diagonally along bottom right. The battle was fought in the marked forest-bordered field north of Holtzwihr.*

On the map above, you can see the inverted U made by the woods north of Holtzwihr. Company A's mortar platoons and a machine gun crew set up at the edge of the woods. What was left of the infantry unit was down to about 400 rounds of rifle ammunition, but some consolation arrived with a pair of tank destroyers that had made their way across the makeshift bridge during the night. As the sun rose, they all awaited orders, wondering what the Germans would do next. From their foxholes, they stared across the snowy field towards the clustered cottages of Holtzwihr, the only sound icy tree branches rattling in the wind.

And that's when one of the most astounding acts of individual bravery in all of World War II occurred. You may have heard of the movie actor and war hero Audie Murphy. A 5-foot-6 Texan all of 19 years old, he had advanced with the 15th Regiment from infantry private in North Africa to second lieutenant in France, having already won Silver Medals for bravery and a Purple Heart from a hip wound. His feat on this day topped all of that.

The Germans waited to attack until 2 o'clock in the afternoon, sending six Panzer tanks onto the field, and behind them two hundred infantrymen in white snow capes, scattered in a skirmish formation. One of the American tank destroyers maneuvered for firing, but slid into a ditch at the edge of the field at an angle that left its turret guns pointing into the air, useless. The engine died and the crew bailed out, retreating into the woods.

Lt. Murphy grabbed a radio and ordered the Company A mortars to begin firing WP rounds to create an obscuring smoke screen ahead of the advancing tanks. Just as he laid down the phone, the first murderous Nazi barrage struck, knocking out the American machine gun nest and the second tank destroyer. The battle seemed lost before it had even begun. Mortar rounds exploded in the middle of the field, though, forming a protective smoke curtain for the moment. Murphy ordered his men to pull back into the woods, but as he turned to retreat, he took another look at the disabled tank destroyer in the ditch. Maybe, he thought, the machine gun in the turret was still usable. Hauling the radio, he climbed aboard, dragged the body of the turret gunner out of the hatch, and took his place, firing a volley at the advancing troops. The rest of this account is excerpted directly from Murphy's memoir *To Hell and Back*:[116]

*Crash! The tank destroyer shudders violently. Vaguely I put two and two together and conclude that the TD has received another direct hit.*

---

[116] Murphy, Audie. *To Hell and Back*. 1949. Grosset & Dunlap: New York.

*The telephone rings.*

*"This is Sergeant Bowes. Are you still alive lieutenant?"*

*"Momentarily." I spread the map on my left palm. "Correct fire: 50 over, and keep the line open."*

*I feed another belt of cartridges into the machine gun and seize the trigger again. The smoke is so thick that I can barely see through it; and the smell of smoldering flesh is again in my nostrils. But when the wind blows the smoke aside, I bore into any object that stirs.*

*The gun has thrown the krauts into confusion. Evidently they cannot locate its position. Later I am told that the burning tank destroyer, loaded with gasoline and ammunition, was expected to blow up any minute. That was why the enemy tanks gave it a wide berth and the infantrymen could not conceive of a man's using it for cover.*

*I do not know about that. For the time being my imagination is gone; and my numbed brain is intent only on destroying. I am conscious only that the smoke and the turret afford a good screen, and that, for the first time in three days, my feet are warm.*

*"Correct fire, battalion, 50 over."*

*The mortar barrage lands within fifty yards of the tank destroyer. The shouting, screaming Germans caught in it are silent now. The enemy tanks, reluctant to advance further without infantry support, lumber back towards Holtzwihr.*

*A dull pain throbs in my right leg. Looking down, I see that the trouser leg is bloody. That does not matter.*

*As if under the influence of some drug, I slide off the tank destroyer and, without once looking back, walk down the road through the forest. If the Germans want to shoot me, let them. I am too weak from fear and exhaustion to care.*

Sporadic fighting continued at wood's edge all day. Lt. Murphy, refusing to be evacuated, taped up his wounded leg and stayed put. His memoir recalls, "That night we lie among the bodies of our comrades who fell at the edge of the forest in the early afternoon." And that's how Lt. Audie Murphy won his Medal of Honor. You may note, on the previous map, a memorial in his name at the edge of the woods where he single-handedly held off that German tank and infantry assault, saving his men – among them my father – from sure death.

The following morning, the Ill River bridges were finally completed, allowing tanks and armor to pour towards Holtzwihr. Their help did not arrive a moment too soon. Some of the rifle companies of Lt. Murphy's 15th Regiment had lost two-thirds of their men in the past four days. Murphy's company, originally comprised of 150 G.I.'s, was down to eighteen men and a single officer, Murphy himself. Lyn's Mortar Company A, which had been laying smoke screens across the open field, switched to rifles, and urged on by squad leader 2nd Lt. Harry R. Freyer, they rounded up a group of scattered infantrymen and moved forward in a desperate effort to hold off another enemy attack. Finally, at sunset the 30th Infantry Regiment, supported by tanks and artillery, relieved the sorely depleted 15th Regiment, taking Holtzwihr the next day.

On January 28th, another three feet of snow fell on the weary troops. Trudging through waist-high drifts in biting cold, the two regiments (joined by the 7th Infantry Regiment) moved on to the last water obstacle before the bridgehead at Neuf-Brisach, the Colmar Canal, a fifty-foot wide ribbon of frigid water flanked by 12-foot high earthen banks. At nightfall, the 99th CMB mortars joined by an entire antiaircraft battalion launched a three-hour barrage across the canal against the fortified town of Bischwihr to clear the way. Then, paddling rubber boats, the G.I.s crossed, engineers shoved over footbridges, and by midnight the infantry regiments had made it to the south side of the canal. Even as snow continued to fall, the weather had warmed slightly, turning icy fields to slush. Mud caked heavy boots and smudged snowsuits. The spoked wheels of the mortar carts sank deep in the ruts. Murphy wrote that uniforms were so ruined that the shape of a helmet was the only way to tell an American from a German soldier.

As the infantry regiments crossed the canal, all companies of the 99th CMB laid smoke screens and placed harassing fire on the towns of Wihr en Plaine, Bischwihr, Fortschwihr, and Muntzenheim. Cold as it was, they were firing so rapidly that their mortar barrels were over-heating. One crewman from each squad had to pack snow around the barrel to keep the propellant charges from igniting before the shells struck the firing pins. All that effort paid off, though. The mortar battalion's online narrative states: "The [3rd Infantry Division] said it was the most successful river crossing they had ever made." Lyn's Company A was lauded for striking a direct hit on an enemy ammunition dump in Muntzenheim. With the German army in disarray from the 16,000-shell artillery barrage, the 3rd Infantry Division captured six towns in just eight hours.

Charging forward over the last two days of January, the 99th CMB supported the infantry attacking Horbourg, Andolsheim and Urschenheim, laying down smoke screens to cover advances over open terrain, as the 3rd Infantry Division closed on its target, the walled town of Neuf-Brisach, while the French First Army, pushing up from the southern edge of the Colmar Pocket, moved on Colmar itself. The 99th CMB narrative summarizes the fighting in January:

> During the first few days of the attack by the 3rd Division, the three companies, A, B, & C, lived in the open, sleeping in dugouts and foxholes in the woods, due to the unavailability of other shelter. In spite of the fact that during this time it was snowing and was bitterly cold and damp, the health of the men did not drop to any great extent…. The morale was extremely

high throughout the drive. The companies fired 17,156 rounds of high explosive and 13,974 rounds of white phosphorous 4.2" mortar ammunition during the month of January 1945. The extremely cold weather caused very heavy breakage of mortar equipment but the untiring efforts of Det. A, 12th Chemical Maintenance Company, enabled the battalion to give vital support at all times.[117]

During the constant battles in Alsace during the month of January, 14,000 Allied soldiers were killed, wounded, captured or missing against an estimated 23,000 German casualties. The 3rd Infantry Division, to which the 99th CMB was attached, alone suffered 4,500 casualties. Sixteen mortarmen were wounded that month, but miraculously, none were killed. And despite living in foxholes in bitter winter weather, none of the mortarmen complained of frostbite or trench foot.

---

[117] https://www.4point2.org/hist-99.htm.

# February 1945

At the beginning of February, the 99th CMB had been in constant combat for two months, most of it spent outdoors braving the coldest, snowiest winter in memory. February would prove to be just as cruel as the previous two months. As one infantry medic recalled, "It was cold, but not quite cold enough to freeze. Rain fell continually and things were in a muddy mess. Most of us were mud from head to foot, unshaven, tired and plagued by recurrent epidemic severe diarrhea… It was miserable to have to jump from one's blankets three or four times a night, hastily put on boots, run outside into the cold and rain and wade through the mud in the dark to the straddle pit. As likely as not the enemy would be shelling the area, and that did not help. [118]

The fighting in February, for the same mountain towns the Americans had captured back in December, must have seemed like some futile trudge through déjà vu all over again. From morning till night, the mortarmen tended their weapons, pulling shells out of crates, and passing them along to the gunners poised at either side of the angled tube, who took turns plopping them into the barrel and sending them whomping out

---

[118] Ambrose, *Citizen Soldiers*, p. 399.

across the battlefield, launching at a 10-shell per minute pace, either laying down WP smoke screens around fortified villages or switching to explosive ordnance to suppress enemy troops and armored vehicles.

The Germans fired back, of course, their mortars and tank cannons seeking to take out American positions. It was tedious, nerve-wracking, exhausting work in weather that alternated between sleet and snow, interrupted only by orders to break down their gear and haul it on to the next town to do it all over again. That was the worst part of the job, because, as Daddy recalled, when you trudged through the village your shells had been falling on all day, you saw the deathly horror you'd caused – mangled corpses of soldiers and civilians, sometimes those of mothers and their babies, along with dead cattle and horses and pigs, all strewn about the smoking rubble of a ruined town.

If there was cheer to be had, it was in knowing that at last it looked like the Allies were winning the war in Europe. News came that the American and English armies had pushed the Nazi's back at the Battle of the Bulge and were preparing to cross into western Germany. The Russian Army was pressing across Germany's eastern border. For all Hitler's fanatical determination, he was running out of men and supplies, while the Allies grew stronger all the time. By February, there were 3 million American troops in Europe, with more landing every day.

Frightening rumors circulated, however, some of them printed on propaganda fliers dropped on Allied lines by German planes. They spoke of a secret weapon, a heretofore unimaginable killing machine that would end the war in an eye blink. The Germans did seem to have come up with terrifying new weapons, the most daunting of them the V-2 rocket, the world's first long-range guided ballistic missile. Traveling at supersonic speeds and hauling a ton of explosives, hundreds of these rockets had been raining down on London, Antwerp, and surrounding cities since mid-November. Because V-2 rockets flew faster than the speed of sound, they hit the ground and exploded, devastating a radius of a full city block, before anyone heard them coming. Rather than use them for close-up fighting on the battlefield, Hitler ordered V-2s aimed at defenseless cities, in retaliation for the relentless Allied bombing that was leveling German cities and armory manufacturers.[119]

And then, one morning, Lyn's mortar squad looked up and saw something that made their jaws drop. This is another story that Daddy told. He said a single German plane came tooling towards them through a mountain pass at low altitude, some suicidal Luftwaffe pilot they all thought, who was daring to risk the bristling weaponry lined up below him. Everybody grabbed their rifles and fired at the plane, which responded by tipping up vertically on its tail and disappearing into the sky, leaving a long exhaust trail in its wake. It was the world's first fighter jet, the Messerschmidt Me-262. How could even the fastest Allied propellor-driven fighters match that thing?

There were other rumors, too, about a bomb that could blow up a whole city, and a rain of propaganda fliers warned that Germany itself had been turned into a mined explosive trap, its civilians fully armed and determined to die rather than surrender to the Allies. But Lyn and his buddies had little time to gossip. On February 4th, they reached their target, the riverfront city Neuf-Brisach, and the 8-pointed star-shaped medieval

---

[119] The classic 1974 novel *Gravity's Rainbow* by the American author Thomas Pynchon depicts what it was like to live in England during the year of V-2 assaults. The book's title notes the arc of a V-2's approach.

ramparts that surrounded it. They bombarded the ramparts for two days, and lay down smoke screens for infantrymen, who were led by civilians through moat tunnels under the city walls. What was left of the German 19th Army had evacuated back into Germany across the Rhine River bridge, blowing up the bridge after the last man crossed, and when the G.I.s entered Neuf-Brisach on February 6th, there was no one left to fight them. 50,000 German troops had made it safely over the river; across the battlefront, 22,000 others left behind surrendered peacefully.

*Mortar and artillery fire on Neuf-Brisach*[120]

---

[120] From https://warfarehistorynetwork.com/article/destruction-of-the-colmar-pocket/

*The gutted nave of St. Louis Cathedral in Neuf-Brisach (later rebuilt)*

Operation Cheerful had proved a resounding success, breaking Hitler's Operation Nordwind assault, smashing the Colmar Pocket, and sending the German 19th Army scurrying back across the Rhine River. Lined up along the river bank, for the next week all of the mortars of the 99th CMB fired on boats, barges, and the town of Vieux-Brisach on the eastern side of the river, which had hitherto escaped Allied bombardment. That's how Lyn spent his third wedding anniversary.[121] The next day, Valentine's Day, mortarmen received orders to pack up and head north. The Germans in the northern reaches of the Vosges Mountain were holding out to a man along the fortified Maginot Line towns that the 99th CMB knew so well. The battalion joined a truck convoy that took them north to St. Avold, a headquarters town near Sarrebourg, where Company A learned

---

[121] The Allied bombing of Dresden, Germany, which obliterated the city in a massive firestorm, occurred that same night. Kurt Vonnegut's famous novel *Slaughterhouse Five* describes its horrifying aftermath.

it was to join the 63rd Infantry Division, while the other two mortar companies were assigned to the 70th Infantry Division, both attached to XV Corps.[122]

Like the 99th CMB, the 63rd Infantry Division has seen its first combat in early December in the Vosges Mountains and had been engaged in tit-for-tat and town-to-town defensive battles along the Saar River for weeks.

*Defensive line of 63rd Infantry Division in mid-February 1945 along the Saar River*

---

[122] From the 99th CMB online history: "The 70th Infantry Division at this time was holding defensive positions along a line running generally from Morsbach, Gaubiving, Bousbach, along the north edge of Bois de Grossblitterstroff, to a point on the Saar River opposite Rilchingen, thence south along the Saar to Welferding. The 63rd Infantry Division was holding defensive positions along the Saar River on the right flank of the 70th Division, on a line running through Saareguemines then north along the Saar to Bliesbruck, and thence southeast along the north edge of Bois (Forest) de Bliesbrucken. Upon first joining the 70th Division, the battalion CP was located in the town of Leyviller, as was also the Medical Detachment. Company A moved into the town of Saareguemines, Company B in Bousbach and Gaubiving; and Company C in Gaubiving, Rouhling, and Morsbach."

With no time to rest, the mortarmen joined in the fighting, and for the balance of the month they supported the infantry in their skirmishes against the German's Saar River defenses. It was the same war they'd been fighting all along, taking one town only to lose it the next day and regrouping to retake it, meanwhile pounding what had been pretty hillside villages into rubble. The weather was miserable, dank and rainy, the creeks and rivers rushing with melting snow, and the farm fields had been churned to bog. Tanks on both sides sought whatever high ground they could find or lurked in the wreckage of cobblestone streets; G.I.s, already burdened by their weapons and backpacks, hauled wooden planks that they lay across ankle deep mud. They covered their foxholes with pine fronds and canvas tarps in a futile effort to stay dry. By this time, any food stored in the ruined mountain towns was gone, the villagers were hungry and frightened, and the G.I.'s survived on tinned rations. Though the German forces had thinned considerably, and infantry squads sometimes were able to take towns without mortar support, the Germans had heavily mined and booby-trapped those towns before retreating, their explosives killing and injuring hundreds of American soldiers.

On February 20,[123] in the center of the just liberated city of Colmar, General Charles de Gaulle, commander of French forces, with great pomp and ceremony honored the 3rd Infantry Division (and its attached units) with his Army's highest honor, the Croix de Guerre with Palm, for their

---

[123] Four days previously, US Marines with massive Navy support, had taken the Japanese fortress island of Iwo Jima.

16-day effort in closing the Colmar Pocket.[124] The 99th CMB, fighting up north, missed the event at which they too received this award. Along with the 3rd Infantry Division, they were also awarded a Presidential Unit Citation for the Colmar battle (excerpted as Appendix C at the back of this book).

Lyn celebrated his 24th birthday in a muddy foxhole in a cold rain, having assisted the 63rd Division in clearing the towns of Kleinbittersdorf, Auersmacher, Bubingen, and the northern edge of the Hinterwald Woods over the previous two weeks.[125] Some days it seemed that the Germans were retreating, their forces growing weaker; other days they fought back as fiercely as ever. Their artillery shells exploded all about, mined roads blew up tanks, machine guns mowed down American G.I.'s as they ventured forward across open fields, and snipers lurked in steeples of conquered towns, just as they had all winterlong. The task now was to clear the Nazi's out of France, pushing them back across the Rhine River into Germany. The task for the Germans, who could no longer resist that relentless shove all along their border, was to kill as many Allied soldiers as they could along the way.

---

[124] The 3rd Infantry Division had already been awarded a Croix de Guerre for its December efforts fighting in the Vosges Mountains, so was also granted the braided French fourragere.

[125] The other two mortar companies, assigned to the 70th Infantry Division, had driven north through Oeting, half of Forbach, Behren les Forbach, Kerbach, Etzling, Spicheren, Lixing, Grossblitterstroff, Zingzing, Alsting, and held the high ground in the forest of St. Arnual.

Allow me to pause for a minute here. Daddy, as noted, turned 24 on the battlefield. Keep that in mind as you read, perhaps reflecting on your own experience at that age. And remember, he had turned 22 in the Texas border country learning how to fire an anti-aircraft cannon, had celebrated his 23rd birthday with U-rations cooked over a tin can stove in North Africa, and by his 24th birthday in an icy trench in France had not seen home in nearly three years.

As testimony to the heavy level of fighting in February, the 99th CMB online narrative lists 9,224 rounds of HE and 13,423 rounds of WP fired, during a month in which the battalion won a Battlefield Unit Citation and the French Croix de Guerre for its key role in breaking the Colmar Pocket, then proceeded without delay to fight entrenched German forces along the Maginot Line. In that short, brutal month, the mortar battalion lost 19 wounded and one killed, while the Seventh Army to which they were attached suffered 7,168 battle and 16,224 nonbattle casualties.[126] The French counted 4,316 battle and 36,540 nonbattle casualties in Alsace, and the German Nineteenth Army listed its losses at 22,000 men killed, wounded, or captured.[127]

---

[126] From Kennett, L., *The American Soldier in WWII*: Postwar calculations showed that the most hazardous roles were combat engineers and medical men, followed by general infantry (264 wounded per thousand per year), then armor, with field artillery trailing considerably (50 wounded per thousand per year).

[127] From Devers.

# March 1945

Eisenhower and his senior commanders felt the same way about the impending conclusion of the European war. By the end of February, it was clear that victory was assured, although no one knew how quickly it would come. The Allied forces in western Europe had grown to 3.7 million men organized in three army groups, nine armies, twenty corps, and seventy-three divisions versus eighty German divisions that looked formidable on paper but were undermanned and constantly harassed by Allied bombers[128]. The Nazis could not reinforce their western front due to a Russian winter offensive that had carried the Red Army to the Oder River in Poland in February. The Allied press from the west and the relentless Russian assault from the east were squeezing Hitler's armies back into Germany, forcing them to defend two borders at once.

Eisenhower planned to launch his final offensive in three phases. The main effort in the first phase was to be carried out by British Field Marshall Montgomery's army group north of the Ruhr Valley, with the objective of

---

[128] This website lists day-by-day combat operations of the American Air Force in European and Pacific Theatres for the entire war:
https://media.defense.gov/2010/May/25/2001330283/-1/-1/0/AFD-100525-035.pdf

getting to the Rhine River. Twelfth Army Group was to protect Montgomery's right flank and then advance to the Rhine in the second phase. The Sixth Army Group, which had rushed to the Rhine at the end of November only to be ordered to hold back, was to remain on the defensive until the third phase, when the Third and Seventh Armies[129] were to capture the German industrial area known as the Saar. Each army group was instructed to prepare plans to execute this strategy.

Having finally claimed the fortified French towns along the Maginot Line that they had been swapping back and forth with the Germans all winter, Allied troops faced a pair of daunting obstacles to this mission. One was the broad, deep and fast-moving Rhine River, swollen as it was with melting Alpine snow. Only a few bridges crossed the river, and the Germans were expected to blow them up after crossing them in any retreat, as they had done at Neuf-Brisach in early February. The other obstacle was Germany's answer to the French Maginot Line.

Germany's West Wall, also called The Siegfried Line, ran along its western border from Holland to Switzerland, a 400-mile long network of 18,000 concrete bunkers, tunnels, pillboxes and tank traps begun in the 1930s as a defensive barrier, and heavily reinforced (by slave laborers) during World War II.[130] If the Allies were to push the Germans back all along their border, as General Eisenhower insisted, then they would need to pry them one skirmish at a time from their Siegfried Line fortifications.

---

[129] The 99th CMB was part of the Seventh Army throughout the War.

[130] Remains of two bunkers with tunnels can be visited today at the West Wall Museum near the German town of Niedersimten: https://www.historyhit.com/locations/westwall-museum/

As the Allied generals prepared for their final assault, the Seventh and French First Armies established a series of rest centers throughout Alsace, providing their combat troops with the opportunity to get out of the line, take shelter from the weather, and enjoy warm showers and food. For the first time since landing in France, Sixth Army Group had enough American divisions to rotate them out of the line periodically. This addressed what General Devers and the army group's psychologists thought was "the main problem of morale and welfare," which was "maintaining units almost constantly in combat." The men of the 99th CMB, however, were not granted a rest break.

At the beginning of the War, Allied psychologists estimated that 60-240 days on the front lines would likely cause soldiers to crack mentally. That wide spread, they claimed, was because some men were tougher than others, and because not all front lines involved constant combat. But during the Normandy campaign after D-Day, doctors saw that the combat effectiveness of troops declined sharply after 30 days of battle, and many G.I.'s, after 45 days, fell into a near vegetative state. 40% of medical discharges during World War II were for what was then called battle fatigue, now known as Post Traumatic Stress Disorder (PTSD).[131] The 99th CMB had been in constant almost daily combat for three months and had lost more than a full company's worth of men, their mortar squads reduced accordingly, yet as their comrades in the 3rd Infantry Division pulled back

---

[131] https:/www.nationalww2museum.org/war/articles/wwii-post-traumatic-stress.

for a well-deserved rest, they were reassigned to infantry units still in combat. Replacements, raw recruits with little training, sometimes arrived to fill out their platoons, including officers unfamiliar with combat operations. The mortarmen did what they could to train up the rookies on the job. No matter the cost to their physical and mental well-being, they had to fight on. There just weren't enough mortar companies to go around.

So, at the beginning of March, Lyn's Company A was still attached to the 253rd Infantry Regiment of the 63rd Infantry Division, which was holding a defensive line at a bend in the Saar River north of Sarreguemines, while the rest of his battalion supported the 70th Infantry Division, further north on the Saar near the town of Forbach.

The 63rd, supported by the Company A mortarmen, had cleared three towns – Klienblittersdorf, Auersmacher, and Bubingen -- in heavy fighting along the Maginot Line during the last two weeks of February, also claiming the northern edge of a thick forest called the Hinterwald Woods. Lyn's Company was now further north than they had ever been, in a war-ravaged triangle bordered by Germany, Luxemburg and Belgium. No longer fighting in Alsace, they had moved into the adjoining district of Lorraine, occupying the western half of the heavily bombarded town of Sarreguemines.[132] German troops occupied the eastern half of the village, with only the narrow Saar River separating the warring battalions.

---

[132] Known for its ceramics and pottery, Sarreguemines (which means "in a bend of the Saar") as it looks today:
https://www.europeanbestdestinations.com/destinations/eden/sarreguemines

*99th CMB crewman Joe Spatola loading a 4.2" mortar during the battle for Forbach, March 1945, in support of 70th Infantry Division.*

During the first week of March, Company A positioned their mortars within the rubble of the ruined town and along the river, but did not fire. Tanks of the 10th Armored Division had joined them and were ranging up and down the river for miles, challenging any German troops who dared to

---

[133] https://www.4point2.org/photogallery.htm.

cross to the west bank. Daring night-time raids across the Saar were ordered, intended to capture prisoners, keep the Germans guessing and test their troop strength. Infantry reinforcements filtered in to replace casualties, and battle-weary veterans were pressed to train the rookies in how to dig in, zero their weapons, and work as a team to take a village. Everyone knew what was coming. As their forces built up along the river, it was like the whole Seventh Army was taking a deep breath, getting ready to pounce on their enemy and push them back into Germany once and for all.

On March 15, after two days of heavy air assault by the XII Tactical Air Command, the Seventh Army, charging with its three corps abreast, smashed at the Siegfried Line. 99th CMB mortar companies A (Lyn's company) and B launched 2360 WP shells, firing from noon until dark laying smoke screens to shield the 63rd Infantry Division's taking of the village of Omersheim. For the next five days, operating like the well-oiled machines they had become, the mortar companies began laying smoke screens at dawn and continued uninterrupted until sundown in the face of German defenses entrenched in the Siegfried Line towns of St. Ingbert and Hassel, and the Saar River city Saarbrucken, which, as the center of Germany's coal mining and steel forging region, had been heavily bombed by the Air Force during the past year. Company C, meanwhile, fired interdicting and harassing fire all day in support of the 70th Infantry Division which was probing for river crossings. Taken together, on just one day of this assault (the 18th), the three mortar companies fired a total of 6,235 WP shells. Yet again on the 20th, Lyn's Company A lay smoke screens from dawn to dusk, alone firing 1,438 rounds of WP.

While researching this story, I came across a fascinating memoir by a man named Rudolph Lea who joined the 99th CMB at this point in the war. It's called *Topsy's GI Journey*,[134] Topsy being Private Lea's pet, a mutt with some terrier in her. Somehow Lea was allowed to keep that dog, which became a mascot of sorts for the battalion, even riding along in the commanding officer's jeep (spoiler alert: Yes, Topsy made it back home to America, safe and sound, at war's end). A raw recruit, a replacement for one of the battalion's many casualties, this is how Lea described his first day on the battlefield, firing on Saarbrucken:

> The platoon had four mortars that were already set up on their heavy metal platforms, and there was a pile of the big 4.2" shells, explosive shells these, not gas shells, stacked on the ground behind them. The men stood around the emplacement, doing small things like making sure the morning dew was wiped off and that all was in readiness. I was wondering what to do when one of the corporals took me to one of the four mortars and just told me to join the men there and help out. I wasn't sure what that meant but I did not ask the corporal, remembering Rule One not to ask questions, especially here, to avoid looking like the replacement rookie I was.
>
> The order to begin firing toward a predetermined target came soon that morning. The squad got busy and one of the guys nodded to me to join him carrying a shell from the pile to the mortar. They were heavy explosive shells weighing 25 pounds, and a technician armed each shell carefully. Then it was lifted by two other guys who, with a practiced movement, dropped it down into the large mortar tube and ducked away, all in one continuous motion, as the shell whooshed out in a high trajectory and with hoped for accuracy toward the Germans. So I became one of four men bringing one shell at a time to the guys firing the mortar.

Lea had been informed about the exploits of the 99th CMB in their four months of nearly constant combat, but he joined them just as that saga

---

[134] Lea, Rudolph. *Topsy's GI Journey: Tales of a Soldier and his Dog in WWII*. iUniverse: Lincoln, NE. 2007.

neared its end. To everyone's surprise, after five long days of relentless bombardment, the German lines seemed to collapse all at once, their troops streaming away in retreat.

The 99th CMB was then sent to join the 100th Infantry Division, which had been fighting near Bitche, France. By this point, the 100th Division, along with the rest of the Seventh Army front, was advancing so rapidly that no elements of the mortar battalion were able to do more than convoy along in case they might be needed.[135]." As the skies cleared, American fighter bombers had a field day destroying German vehicles along roads of retreat. Every foot of those roads was clogged with wrecked vehicles and equipment. One American general said, "It is difficult to describe the destruction. Scarcely a manmade thing exists in our wake; it is even difficult to find buildings suitable for command posts: this is the scorched earth."

On the first day of spring, the 99th CMB crossed into Germany. As Private Lea recalled,

> The sun was shining brightly that day as we descended from high ground, driving slowly down a road clogged with abandoned trucks, wagons, and numerous dead horses. It was a scene of heaps of equipment left behind by a retreating army, made to look almost familiar like an epic period-piece war movie because of the dead animals and debris all over the sides of the road. We saw no human casualties nor any civilians on that very first stretch of road on German soil. German soldiers must have gotten out of there in time to regroup somewhere else. The sunny weather and the emerging beauty of the landscape ahead contrasted strangely with the scenes on the road.

---

[135] From the 99th CMB online narrative: "The battalion convoyed from Sarreguemines to Bitche, France, thence along a route following through Schweyen, Germany, Zweibrucken, Thalmischweiler, Waldfischbach, Speyerbrunn, Waldeingen, Frankenstein, Bad-Durkheim, and Friedelsheim."

*Dale Bell, driver for 99th CMB Commanding Officer*[136]

During the third week of March, as the German forces fell apart, the Seventh and Third Armies captured more than 90,000 soldiers and killed thousands more. Then on the night of March 23, Patton's Third Army crossed the Rhine River near Oppenheim and established a solid bridgehead against light opposition. The Seventh Army cleared the remaining pockets of resistance west of the Rhine and began moving boat and bridge units towards the river along roads littered with abandoned weaponry. On the 25th, the 99th CMB, traveling with the 100th Infantry Division, was halted at the pretty vineyard-framed village of Meckinheim (near Neustadt), nearly 20 miles short of the river, and for the next four days were granted a program of "renovation, rehabilitation, and training," while Rhine River crossings were being built and consolidated. This was the first time that all elements of the battalion had taken a break since commitment to battle on December 2, after 113 days of continuous combat.

---

[136] From *Topsy's GI Journey*, p. 90. Photo by book's author Pvt. Rudolph Lea.

The next day, it would have been in Meckinheim or Neustadt, Lyn yanked a beach towel-sized Nazi flag down from a pole, trimmed it to just the swastika, so it would fit in his backpack, and asked his platoon-mates, his best friends in all the world, to sign it, along with their hometown addresses. 28 did, in black ink that is mostly still legible all these decades later. Though one or two corporals and PFC's listed their rank on the flag, Lyn's three sergeants – Staff Sgt. John Miksch, Staff Sgt. Evert Nelson, and Sgt. Arlie Boatright, Jr. -- did not. Among the signees, by war's end, one – Pfc. Vernon K. Perry would be awarded both a Silver Star and a Bronze Star (along with a Purple Heart); eight others won Bronze Stars, and four earned Purple Hearts.

Pulling down an enemy flag is the classic symbol of victory in any war (or war game). Defacing that flag by cutting it up and scrawling the names of some of the victors on it must have been especially delicious. And that done, the mortarmen wandered the town, good German beer and white Reisling wine in their bellies, imagining maybe for the first time that yes, they might survive this war after all.

*The flag in my study, signed by 28 members of the 99th CMB.
See Appendix B for list of names on the flag.*

*Note 4.2 mortar in place, stack of shells, and damaged building. Guessing these two photos were taken the same day as the shot of three friends in winter whites, as building looks the same. Photos sent home by Lyn Gentry.*

None of it had been easy. Consider this assessment by Brigadier General Hugh W. Rowan, the U.S. Army in Europe's Chief Chemical Officer:

> The work of the battalions on the Siegfried Line is an epic in modern military history, of which the Chemical Warfare Service will be forever proud. These latter operations lack the color and dash of the Normandy Campaign, and hence their story may hold less interest for the reader. To the military student, however, the accomplishments of the chemical battalions during the past four months have been of pre-eminent interest.
>
> So few 4.2 battalions are available, and they are considered so indispensable by corps and division commanders that they have been kept constantly in action, without any rest. For months at a time, they have endured every possible hardship of severe weather, difficult terrain, and enemy action. They have also been called upon to execute missions for which the mortar was never designed, and these missions, too, have been successfully executed. [137]

On March 29, the 99th CMB crossed the Rhine River in line behind the storied 4th Infantry Division. The 8th Regiment of this division had been the first unit to land on Utah Beach on D-Day and two months later, after heavy fighting in Normandy, the division liberated Paris. Novelist Ernest Hemingway had entertained its troops there, and future novelist J. D. Salinger served in the division's 12th Regiment. They crossed the river – swollen with Alpine snow melt, swift-moving and four football fields wide – on a recently laid pontoon bridge that night, into the city of Worms, which had suffered air bombardment and a house-to-house infantry battle just that week. Worms dates back to Celtic times; from the Middle Ages, until Hitler, the city had been a cultural center of Jewish life in Europe. Its

---

[137] https://www.4point2.org/MortarsinNormandy.pdf.

citizens proudly call it the oldest city in Germany, but now it lay in ruins. One G.I. wrote:

> While passing through the city itself [we had our] first glimpse of the damage that could be wrought upon a large city by the Air Force. Hardly a house in the entire city possessed a roof, rubble was piled high in the streets and civilians were busy looting a huge wine cellar. Whole blocks of buildings were completely leveled. All these sights were to become familiar to every man in the Division before the month of April, with its marching and fighting from one large city to another.[138]

Lyn and his mates followed along with the infantry's rapid drive to the east through farmland untouched by war and through the towns of Dienstadt, Schneeburg, Tauberbischofsheim, Olfen, Kailbach, and Ebersbach. Though attached to assault battalions primed for combat, they met only scattered pockets of resistance and did not need to set up their mortars.

In just two weeks, the Seventh Army had broken through the Siegfried Line, crossed into Germany, cleared the west bank of the Rhine, and established a large bridgehead on the east bank. The 99th CMB online narrative tersely summarizes the mortar teams' work as follows: "Seven enlisted men were battle casualties during the month of March 1945, with one resulting in death. 11,887 rounds of HE and 14,709 rounds of WP, 4.2" mortar ammunition, were fired during the month."

---

[138] The website *Critical Past* is a collection of short videos, many from WWII. This link shows clips from the battle for Worms:
https://www.criticalpast.com/stock-footage-video/Worms+Germany+1945_2

# April 1945

Springtime (not so much for Hitler) in Germany. Violets and myrtle everywhere in bloom and fruit trees stippled with buds. Ahead of the millions of Allied soldiers streaming across the Rhine River on improvised pontoon bridges, sappers checked for mines and engineers carted rubble from bombed cities to fill roadside ditches, widening narrow country lanes into two-way roads for what must have seemed like an endless dusty line of trucks, tanks, and jeeps.

As you might imagine, there were traffic jams and mix-ups. For example, on April 2nd, Technician 5th Grade Clarence E. Dixon of 99th CMB Company C, while driving a one and a half-ton truck through heavy traffic in Kutzbrunn, unwittingly found himself in the enemy-held town of Messelhausen. Realizing his predicament, he calmly drove on to the main intersection of town, then whipped his truck around and got to the outskirts before being stopped by enemy small arms fire. He and his assistant PFC William R. Batte shot back and pinned the enemy down long enough to escape to the safety of their platoon. During the encounter, a "panzerfaust" anti-tank shell struck the fender of the truck, which was heavily laden with explosive ordnance, but somehow the shell failed to explode. Even though the radiator of the truck and their sleeping rolls were riddled with bullet holes, both miraculously escaped serious injury.

The 99th CMB was driving east with the 4th Infantry Division, towards Nuremburg, the cradle of the Nazi party, at a 30 mile a day pace, when they were ordered to join the 42nd Infantry Division to assist in the capture of Wurzburg. What they found there was jaw-dropping devastation. Wurzburg was not a manufacturing hub – it was a healing center with forty hospitals – but two weeks earlier Royal Air Force Lancaster bombers had dropped 1200 tons of incendiary bombs, creating a firestorm that killed an estimated 5,000 people and almost completely obliterated the historic city.[139]

That did not stop the Germans from resisting land assault. In the first skirmish American forces had faced since crossing the Rhine River, Nazi artillery on the right bank of the Main River traded shelling with the 99th CMB's mortars, positioned in the courtyard of the Marienberg fortress, an old castle complex on a promontory overlooking the river. The mortarmen were set up in a perfect spot to lob shells at German strongpoints in the city below. Infantrymen headed towards the three bridges across the river, only to see them blown up one-by-one as they approached. Resorting to light boats and a quickly constructed pontoon bridge, G.I.'s entered the ruined city only to face Panzer tanks and a crossfire of snipers. The first platoon of Lyn's Company A joined in the assault, deploying with machine guns as infantrymen during the night-time advance. They captured 17 Nazi soldiers, including a captain, and killed a civilian who was leading a team of Wehrmacht soldiers towards their

---

[139] Almost 90% of Wurzburg's buildings were destroyed in the raid.

position. For three days, the warring troops skirmished in the rubble, until the Germans ran out of ammunition, surrendering what was left of the bombed-out city on April 6th.

Replacement mortarman Rudolph Lea remembered:

> We moved down, crossed the river into the badly damaged city, and drove through streets strewn with the dead. It was difficult not to be struck by the terrible waste of life and destruction in the senseless defense of Wurzburg. In spite of much destruction, we could see what a beautiful city Wurzburg must have been, with its large stately buildings, its famous Residenz baroque palace and its university, all of them badly damaged but still standing. Bombing raids and mortar firings tended to leave walls of buildings standing upright with their insides totally gutted to be sure, whereas artillery shelling would have left them leveled.[140]

A bright spot in all the carnage was the discovery of caves of excellent wine and champagne, which made a feast of the soldiers' C ration meals. The Seventh Army was now in the heart of Germany's famous vineyard region, and for the next couple of weeks, every scouting party brought back crates of wine.

Turning north into Bavaria, the 42nd Division aimed for the ball bearing factories at Schweinfurt, which had been targeted in two notoriously unsuccessful bombing raids earlier in the war.[141] In August 1943, at a time when the Luftwaffe had all the fighter planes and antiaircraft artillery it needed, 60 American B-17s and B-24s were shot down, and in a second raid two months later, 291 American planes, unprotected by fighter

---

[140] *Topy's GI Journey*, p. 98.

[141] Low friction ball bearings were key parts of nearly all German war machinery, and nearly half of the ball bearings produced in Germany were made in Schweinfurt.

*4.2 mortarmen with 3rd Infantry Division, from company newsletter.*

escorts (which had reached their fuel tank limits many miles earlier), were swarmed by German JU-88 fighter planes, which harassed them all the way back to the English Channel. Though some of the bombers got through, damaging the ball bearing factories, 77 B-17s and 600 crewmen were lost on a day remembered by the Air Force as "Black Thursday."[143] Since then, Schweinfurt had been bombed twenty more times, most recently just the day before the infantry arrived, on April 9th. The soldiers had heard the low

---

[142] Image from: https://digicom.bpl.lib.me.us/cgi/viewcontent.cgi?article=1066&context=ww_reg_his.

[143] A recounting of the Black Thursday battle: https://www.nationalww2museum.org/war/articles/black-thursday-october-14-1943-second-schweinfurt-bombing-raid

drone of the bombers and watched the sky etched for an hour with their vapor trails high in the air. As at Wurzburg, the Nazi defenders were armed with 88-mm cannons, and used them, but they surrendered before the day was out. Half the city's houses and most of its factories lay in ruins.[144]

After taking Schweinfurt, the 42nd Division was ordered to turn around and dash 15 miles due south, setting up astride the Wurzburg-Nuremberg Road, in preparation for an attack on the city of Nuremberg. Resistance was so light, and the situation so fluid, that the 99th CMB trucks on several occasions found themselves ahead of the infantry. In one instance, told to join an assault on a fortified town, the mortarmen took a short cut and arrived to find that the Germans had already evacuated, a fortunate thing as the infantry units did not arrive until hours later.

On April 11th, Company C peeled off from the rest of the mortar battalion and rejoined the 4th Infantry Division, which was moving eastward on the right flank of the 42nd Division. Coming upon two companies of Nazi soldiers in a forest, the mortarmen hastily set up their guns for battle. Rather than fight, the Germans retreated, but – probably young and inexperienced – they did so in parade-like formation, making them sitting ducks for Company C's mortars. The 99th CMB narrative states: "Our already zeroed-in mortars literally plastered the road, scoring many direct hits on the column, almost completely destroying the entire group." Survivors of the shelling took cover on a wooded hill, however, and during the night were able to evacuate their wounded and some of their dead. The Americans counted 64 bodies left behind the next day. While Company C

---

[144] This website displays a series of then and now photographs highlighting the devastation in Schweinfurt: http://www.thirdreichruins.com/schweinfurt.htm.

fought in the woods, the rest of the 99th CMB continued on towards Nuremburg with the 42nd Infantry, pausing for a solemn ceremony and salute on April 13th, upon learning that President Franklin D. Roosevelt had died of a stroke.

That news came as a shock to everyone. Having served just short of four terms in office, FDR was the only President most of the troops could remember; he was a hero for all the jobs he had created during the Great Depression and, as commander-in-chief, he was trusted as the true leader of America's war effort. As author Rick Atkinson noted, FDR was "a man who had never fired a shot in anger and yet became the greatest soldier in the most devastating war in history. He had crossed over at the fag end of an existential struggle that would be won in part because of his ability to persuade other men to die for a transcendent cause."[145]

British Prime Minister Winston Churchill sobbed when he heard the news, saying that his American partner "altered decisively and permanently the social axis, the moral axis, of mankind by involving the New World inexorably and irrevocably in the fortunes of the Old. His life must therefore be regarded as one of the most commanding events in human destiny." And Private Rudolph Lea of the 99th CMB recalled that, among the troops: "It felt like the death of a father. I watched two big burly G.I.'s stop in their tracks by the side of the road and cry like babies. I heard no jokes, no banter, nothing but silence and incredulity."[146]

The 99th CMB continued in convoy with the 42nd Infantry Division advance towards Nuremburg, notorious for its extravagant Nazi rallies held

---

[145] Atkinson, Rick. *The Guns at Last Light*, p. 594.

[146] Lea, Rudolph. *Topsy's GI Journey*, p. 105.

in the lead-up to World War II. By this point, Hitler had ordered a fight to the death (deserters were being shot without trial). Civilians in the surrounding cities were issued steel helmets and rifles and ordered to resist invaders at all costs. Hitler also issued a desperate "Nero Decree" ordering the Army to booby-trap every city with explosives armed to self-destruct as Allied forces arrived.

When the 42nd Infantry reached Furth, a city just ten miles from Nuremberg, they ran into a hailstorm of defensive fire. The 99th CMB had been ordered to leave the 42nd and join the 63rd Infantry Division's march, but the fighting was so intense at Furth that Companies A and B were called back to provide smoke-screen and explosive support there. Though outmanned and outgunned, Nazi troops in Furth put up a fanatical resistance in an effort to keep the Allies out of Nuremberg. Every weapon to hand came into play: mortars, tanks, machine guns, bazookas, grenades, even pitch forks, as the Germans fired from ready-made dug-outs in bomb craters and set up as snipers, it seemed, in every broken window. Civilians threw handheld panzerfaust anti-tank shells, mined streets exploded, and the G.I.'s found themselves skirmishing for hours just to claim a city block.

Finally having taken Furth on April 18th, the 42nd Infantry Division pushed on towards Nuremberg, while Lyn's mortar battalion was ordered to make a 60-mile dash south to join the 63rd Infantry Division, which was seeking a secure crossing of the Danube River near Gunzburg. Catching up with the 63rd Division near Langenberg, the mortarmen were placed under control of the division's artillery command. Despite road blocks, mines and blown bridges, they all made good time, pausing occasionally to set up their mortars and fire on enemy road blocks. They seized bridgeheads at the Danube on April 25th, Company C firing 816 rounds of WP to screen the bridge as they crossed. Everyone ducked for cover when a pair of German twin-engine Messerschmidt 282 fighter jets strafed the road. Rudolph Lea noted: "The 99th escaped casualties; we must have had a charmed existence because evidently most of the bullets landed on the houses and in the fields."[147]

Earlier that month, on April 1st, another German jet had strafed Company A. Corporal Byron Lemmon, who had been awarded a Bronze Star and Purple Heart while fighting for the Colmar Pocket in February, ducked for cover in a nearby building. Years later he recalled, "A sniper from somewhere nearby took a shot at my head but hit me in the shoulder instead. So I was evacuated by jeep to the 4th Division aid station and then loaded into an ambulance to try and find a hospital."[148] Combat ended for Corporal Lemmon that day, his injury occurring just four days after he signed Lyn's swastika flag. Fortunately, these were some of the last sorties by the German jets. Luftwaffe airstrips by then had been cratered by bombing runs, fuel was running low, and ranks of the weird, propellerless planes squatted with broken wings on Autobahn highways.

Deep in Bavaria now, the G.I.'s marveled at how pristine these villages looked compared to the 85 battle-ruined French towns Lyn's battalion had shelled amidst all the give and take of the winter's battles. The large German cities, especially those where wartime manufacturing was centered, had been shattered by four years of Allied bombardment, but the smaller towns seemed entirely untouched by war. The convoy wended southeast through some of the prettiest countryside in Europe, where yes,

---

[147] Lea, Rudolph. *Topsy's GI Journey*, p. 103.

[148] https://www.dvidshub.net/news/129438/world-war-ii-veteran-visits-former-unit.

the German citizens had suffered rationing, where the young men had been pulled into the army, but where the townspeople seemed to be getting along quite nicely, thank you.

The G.I.'s billeted in comfortably furnished homes with full larders and linen closets, furnished kitchens, and bathrooms with plumbing, a far cry from the barns and icy foxholes they'd known in France. Every town had its butcher shops, bakeries, grocery stores, and wine shops, and the troops took what they needed, as invading armies always have. The German people were not happy to find their sleepy towns overtaken, of course, but having no choice in the matter, coped as best they could. As Stephen Ambrose writes in *Band of Brothers*:

> The average G.I. found that the people he liked best, identified most closely with, enjoyed being with, were the Germans. Clean, hard-working, disciplined, educated, middle-class in their tastes and lifestyles (many G.I.s noted that so far as they could tell the only people in the world who regarded a flush toilet and soft white toilet paper as a necessity were the Germans and the Americans), the Germans seemed to many American soldiers as 'just like us'".

For Lyn, who had grown up using an outhouse toilet and bathing in a kitchen wash tub, those quaint village homes must have seemed palatial.

The great joke amongst the troops, however, was that no one in Bavaria owned up to their part in the war. Everyone claimed to hate Hitler. Journalist Martha Gellhorn (a brilliant war correspondent married to Ernest Hemingway at the time), wryly noted:

> No one is a Nazi. No one ever was. Oh, the Jews? Well, there weren't really many Jews in this neighborhood. Two maybe, maybe six. We have nothing against the Jews; we always got on well with them. We have waited for the Americans a long time. You came and liberated us. The Nazis are Schweinhunde. No,

> I have no relatives in the Army. I was never in the Army. I worked the land.

Gellhorn added of their plaints, "It should be set to music," because, at night:

> The Germans take pot shots at Americans, or string wires across roads, which is apt to be fatal to men driving jeeps, or they burn the houses of Germans who accept posts in our Military Government, or they booby-trap ammunition dumps or motorcycles or anything that is likely to be touched. But that is at night. In the daytime we are the answer to the German prayer, according to them.[149]

At noon on April 25th, orders arrived to split the three companies of the 99th CMB. Company B was relieved from attachment to the 63rd Division and joined the 3rd Infantry Division then preparing to assault Augsburg (the city fell on April 28th). Lyn's Company A attached to the 36th Infantry Division near Weilheim, and Company B went to Wolfratshausen to join the 4th Infantry Division. The battalion's online narrative summarizes their work in April as follows:

> The morale of the men continued high, in spite of the almost ceaseless moving. Meals were often postponed and night moves were not infrequent. Convoys were occasionally strafed by single enemy aircraft but no casualties or damages were sustained.
>
> One factor contributing greatly to the morale of the unit was the fact that they were part of a great offensive Army, the American Seventh, which was moving relentlessly forward against the enemy. They were biting deep into the very cradle of Nazidom, giving the disorganized, retreating enemy no rest and denying him the opportunity to establish any last line of defense for a "National Redoubt" in the

---

[149] Gellhorn, Martha. *The Face of War*. Grove Press: New York. pp. 239-40.

Bavarian Alps. The men could see that the collapse of the enemy was imminent and final victory was in sight.

Eight enlisted men were battle casualties during the month of April, with one of these resulting in death. During the month, 3,115 rounds of HE and 3,342 rounds of WP were fired.[150]

---

[150] https://www.4point2.org/hist-99.htm.

# May 1945

It happened on April 30, but the news broke the next day. Nazi dictator Adolph Hitler was dead. At first, German propogandists claimed he went down in a hail of gunfire, defending his Berlin headquarters, but the truth soon came out. Hitler had put a bullet into his own head. Which meant, of course, that the European War would soon be over. This news may have arrived as cold comfort to Lyn and his comrades, however, because, also on the last day of April, they came upon and liberated one of the Kaufering subcamps of the Dachau concentration camp system near Landsberg.[151] (The 101st Airborne Division joined in liberating Kaufering, as shown in the film *Band of Brothers*). The Kaufering subcamps were some of the worst outposts in the enormous German concentration camp system. Most of the captives were Hungarian Jews, shipped there to work on starvation rations constructing a vast underground bunker, where the Nazi's hoped to build more of their fighter jets safe from Allied bombing raids. The captives had slept on vermin-infested straw in holes in the ground covered with thatched roofs that provided little cover from the

---

[151] A harrowing seven-minute long video discussing what the G.I.'s found at Kaufering: https://www.youtube.com/watch?v=rgAmDi96qpg.

weather. Ahead of their retreat, the German guards bolted shut the doors to those dwellings, doused them with gasoline and set them afire, burning alive the 4,000 captives inside, and leaving the G.I.'s to discover rows of charred and still smoking corpses, naked bodies piled in stacks, and a few living skeletons, who came to them begging for help. One of the liberators recalled:

> Our first sight of the camp was appalling. Inside the enclosure we could see three rows of bodies, approximately 200, mostly nude. We entered the camp to look it over. The bodies were in all shapes and conditions. Some were half burned, others badly scorched. Their fists were clenched in the agonies of their death. Their eyes were bulging and dilated as though even in death they were seeing and enduring the horrors of their lives in prison. None were more than skin and bones.[152]

Rudolph Lea remembered:

> …for lengths of an entire city block, we saw piles of bodies about two to four layers high. At first I sought for a way not to look, to escape my personal horror and my fear of the horror. But there was no escape. I stared at the immense long pile. I looked around me as if to seek help from others, but my friends, and any others present, all stared in the same way and in total silence.
>
> We saw no survivors, nor learned if any emaciated bodies had been found still breathing. Around the center of the immense pile, all the bodies were blackened. They bore witness to the terrible atrocity committed by the SS Guards just before the camp's capture, when they poured gasoline over any living inmates who remained and the large numbers of dead and set all ablaze to destroy the evidence. They obviously had failed badly.[153]

---

[152] Blatman, Daniel (2011). *The Death Marches*. Harvard: Harvard University Press.

[153] Rudolph Lea. *Topsy's GI Journey*, p. 109.

Writing this, I can see my father now, as I described him in the preface, trembling before the tv he had just snapped off in anger at the claims of a late-night Holocaust denier. A quarter of a century on from this horror, exclaiming: "He's a liar! I saw bodies stacked from here to Fork Union!" Clearly, it was all still fresh in his memory, seared there, even if he never spoke of it again.

The 99th CMB, less Company B, settled within sight of the Bavarian Alps along with the 36th Infantry Division, Company A billeting in the picture postcard-pretty village of Seeshausen, which sits on sky blue and island-dotted Lake Staffelsee. The whole Seventh Army by now had converged on southern Germany, because of a threat that some of the Nazi's would hole up in the mountains to fight a guerilla war to the death there. That threat never materialized. On May 2nd, after a week of bombardment, in which the Soviet Red Army fired two million artillery shells on Berlin, the tattered remnants of the German army there surrendered. Four days later, an end to all combat in Europe was declared, and on Mama's 24th birthday, May 7, the German generals surrendered. On the following day, known as Victory in Europe Day (VE Day), formal surrender documents were signed. After 150 days of active combat, the 99th CMB's fighting was done.

The war in Europe may have been over, but the catastrophe that the war left behind had to be managed. Charged with security missions and patrol activities, the men of the 99th CMB must have stood agape at the endless lines of lost and wandering people on every road. Among them were 90,000 Allied soldiers and airmen held as prisoners of war in Germany and its satellite states, who had been summarily freed to find their way back to their comrades.

In April alone, 2 million German soldiers in tattered uniforms, but still in strict formation, had surrendered. Slave laborers from all over eastern Europe, who had been shipped to Germany to serve the Nazi war effort, stumbled along like skeletal zombies in their blue and white striped prison garb (it is estimated that 250,000 were liberated from concentration camps at the end of the war, despite concerted efforts by the Nazi's to kill them all first). Thousands of urban residents sat stunned in the ruins of their homes.

The seemingly endless tide of refugees on every cart path carried tuberculosis, typhoid, diphtheria and other diseases. Some died as they marched and were buried in shallow graves in roadside ditches. Those liberated from slave camps sometimes rioted, killing their former guards, plundering houses, burning furniture for campfires, and raiding breweries and grocers with such ferocity that they sometimes died from over-indulging their shrunken stomachs. For the men of the Seventh Army charged with keeping order, it was all too much. News correspondent Eric Severeid tried to sum up his feelings amidst the post-war tumult: "a kind of dull satisfaction, a weary incapacity for further stimulation, a desire to go home and not have to think about it anymore – and a vague wondering whether I could ever cease thinking about it as long as I lived."[154]

Lice-infested prisoners were dusted with DDT (DDT was the standard de-louser for Allied soldiers throughout the war) and given clothes pulled from the closets of German civilians. Wooden barracks were hastily thrown together to house the refugees. Everyone needed to eat, to have shelter, to find their scattered loved ones and their way home; among them, the millions of Allied soldiers and Soviet Red Army troops now crowding

---

[154] Atkinson, Rick. *The Guns at Last Light*, p. 600.

into Germany. This was a tactical and logistical problem no less difficult to solve than the war's battle plans, and the men of the 99th CMB were tasked with pitching in to support that effort.

Which brings me to my favorite photograph in this book (next page). Please pause a moment to take it in. That's 23-year old Army private Lyn Gentry, in torn slacks but wearing his cap at a jaunty angle, smiling proudly, in his arms a refugee child, perhaps recently liberated from a concentration camp. Behind them, a barracks kitchen, its screen door open wide to let in the spring breeze, cook pans hung to dry on the clapboard wall. This young soldier had the photograph developed and sent home to a bride he had not seen in nearly three years.

Think of what he meant to say with that photograph, what it must have meant to her. Amidst all of their security and assistance work, the troops found time in early May for raiding wine cellars, swimming in icy Alpine lakes, relaxing in flowery meadows admiring the snow-covered Bavarian Alps arrayed along the southern horizon, and snatching souvenirs to take home. G.I.'s mailed silverware services, mink coats, even motorcycles commandeered from their German owners.[155] Lyn brought home, along with his autographed swastika flag, a finely engraved antique dueling pistol, taken from a university display case somewhere in Germany. My sister Kay has it now, intending to pass it along to a grandson.

---

[155] One enterprising radioman J. Herbert Orr, discovered a newfangled device called a Megnetophon that used reels of tape to record audio. He shipped two of them home to California, and upon his return to the States copied them, launching a revolution in musical recording.

The biggest topic of discussion among the Allied troops at this point was whether they would get to go home and take off their uniforms or be shipped to Japan to continue fighting there. The Army had configured a points system that required 85 points for demobilization,[156] and every G.I. ran the numbers to see where he stood. The old-timers of the 99th CMB, who had been with the battalion since Texas, even those with children back home, came up short. My calculation for Daddy, adding in 5 points for the battalion's Presidential Unit Citation (which may not have applied, since points may have been allowed only for individual medals) amounts to 55. These men had been in almost constant combat under grueling conditions for 150 days of the European campaign. But once their clean-up work in Germany was done, they were all going to Japan.

Until mid-May the battalion, per the online narrative, continued their patrols "searching towns and woods, collecting and controlling released

---

[156] Points were awarded as follows: Each month in service = 1 point (since September 16, 1940); each month in overseas service = 1 point (additional); any combat award = 5 points; each dependent child = 12 points. Strangely, no points were awarded for days in combat.

Russians, Czechs, Poles, and other nationals, and maintaining order." This was an enormous challenge, as there had been 170 slave-labor camps dotted about the pretty Bavarian countryside, all of which were liberated by the Seventh Army.

On May 16th, the men of the 99th CMB took one last look at the mountains reflected in Lake Staffelsee before climbing into their trucks and heading north, stopping for the rest of the month in Nussloch (literally "nut hole"), not far from where they had crossed the Rhine River two months prior.[157] It would take two more months to get there, but (even if for only a short respite ahead of deployment to Japan), the 99th CMB was going home.

The Battalion's online summary reads: "There were no battle casualties during the month of May 1945. Only 62 rounds of HE and 42 rounds of WP were fired by the battalion during the month. This brought the total number of rounds expended by the battalion during its commitment to combat from 2 December 1944 to 6 May 1945 to 56,662 rounds of HE, and 56,380 rounds of WP, or a grand total of 113,002 rounds."

The online narrative does not summarize casualties during the war, but in the Battalion as a whole, which arrived on the front with 33 officers, 1 warrant officer, and 556 enlisted men, final casualty numbers amounted to 71 wounded and 5 killed. One more, an officer, would die before they reached home.

---

[157] The route of march was Holzkerchen, Munich, Augsburg, Donauworth, Nordlingen, Ellwangen, Schwabisch Hall, Lowenstein, Heilbronn, Heidelberg, and Nussloch.

# June 1945

For two weeks, the 99th CMB billeted in Nussloch, awaiting their turn for a ship back to the U.S. No longer charged with policing, they spent their time repairing their much-abused weapons and machinery. Rudolph Lea recalled: "We did a lot of polishing in Nussloch; carbines were disassembled, cleaned, and oiled, mortars restored to top condition, jeeps and trucks cured of their aches and pains."[158] Nussloch was a typical German small-town, spared from combat, but in addition to the soldiers, it was crowded with refugees from the nearby city of Mannheim, which had been demolished by bombing raids. Lyn and his buddies were surprised to see that the citizenry were already putting the city back together. Piles of rubble stood at the end of every street, trolley cars were up and running again, and hawkers sold spring strawberries and flowers in makeshift outdoor markets.

Every soldier in every war wonders who will be the last to die. In the first week of June that question was answered for the 99th CMB, when Company B's 1st Lieutenant Victor M. Harris' jeep hit a rut, ran off the road

---

[158] *Topsy's GI Journey*, p. 130.

and crushed him against a tree. Lt. Harris had led a platoon of Company B since their training days in Texas and had survived every battle of the war, only to die on the way home. Rudolph Lea recalled, "He was a model officer and he had a gracious mind and he symbolized the best qualities of the Battalion. Everyone admired him and we mourned his loss."[159]

On June 14th, a gorgeous spring day, excitement ran through the camp, when the men received orders to turn in their weapons and prepare to ship out. All but one platoon of Company A departed by truck convoy, along with the Headquarters and Medical Companies. In nearby Heidelberg, the rest of the battalion boarded 40 x 8 railcars, like the ones they'd ridden on the way to the battlefront seven long months ago, en route to Valery en Caux, France, and the "Lucky Strike" staging area. No longer attached to the Seventh Army, they were now under the control of Le Havre Port of Embarkation.

Along the way they retraced the battlefields they knew so well, crossing into France at Saarbrucken and getting one last long look at the destruction the war had wrought on that ancient riverfront city. Back in Alsace, the next town they came to was Forbach, where Company B had fought so hard in the February blizzards. They slept in French barracks that first night outside Metz, a walled city that Patton's Third Army tanks had pummeled, then rejoined the long line of truck convoys and troop trains inching west towards the French coast.

The next day they passed through Verdun, noting the cemetery where 130,000 unknown soldiers from the longest battle of World War I are buried, then paused in Reims, so everyone could grab a bottle of good

---

[159] *Topsy's GI Journey*, p. 139.

champagne for their second overnight stop, a bivouac in pup tents near Soissons, another battleground of the First World War.

The 99th CMB never saw Paris. Their truck and rail convoys passed 50 miles north of the French capital through Compeigne, where the Armistice to end World War I had been signed and where the French had surrendered to the Germans in 1940, finally reaching Camp Lucky Strike near the port of Le Havre in the late afternoon of June 16th. They were in Normandy now, on the wreckage-littered coast where the D-Day invaders had landed a year and a week ago. They checked into barracks in the town of St. Valery en Caux, stashed their personal items in newly issued barracks bags, and then, as with so much in the Army, they waited. For two more weeks, they waited. One antsy soldier recalled:

> During our…stay at this camp, we had a chance to visit the city of Le Havre. It was about a 30-minute walk into town. We stayed a couple of hours, but there was not much to see and very little to buy in the stores. It was so soon after the war ended that the stores were not re-stocked. I would like to see the change that must be there today. I do remember a small carnival operating there in the town and I think we rode the merry-go-round.

Then at dawn on June 29th, everyone mustered into trucks for the short ride to the port, where they climbed aboard open-mouthed amphibious landing vehicles parked side-by-side along the beach. On D-Day, thousands of Allied troops had poured out of LCT's like this along the Normandy Coast. Filing onboard now looked like a film run backwards, minus, of course, that murderous Nazi shelling. Rudolph Lea recalled:

> The LCT's ramp came up, the big mouth closed noisily and then it backed away from the shore, reversed its engines, and angled out into the harbor, chugged around a small headland,

and then headed straight for the largest vessel we could now see dimly, in camouflage gray, anchored some distance ahead. We were in great spirits as we felt the long-awaited sensation of actually beginning the trip home.[160]

The USS Wakefield, a troop transport ship during the war, in civilian life had been the luxury liner SS Manhattan, the fastest passenger ship in the world. Enormous at a sleekly stream-lined 705-feet long, the ship was powered by giant steam turbines driving two screws. She could outrun a U-boat, hauling 7,000 troops at a time on her Boston-to-France runs.

*U.S.S. Wakefield loaded with troops.*

Coming aboard, each soldier was issued an assignment card indicating the location of his bunk and dining schedule; on the back directions for fire drills and how to abandon ship were listed in small type. By noon the ship was loaded but sat at anchor that night. Tugboats nudged the Wakefield from the mouth of the Seine River as the sun rose on the last

---

[160] *Topsy's GI Journey*, p. 142.

day of June 1945, and she steamed out across the Atlantic Ocean, destination Boston, USA.

Onboard ship, the troops were served two meals a day, which they ate standing up in four chow lines. A thousand men were fed every 20 minutes out of steaming kettles filled with oatmeal and dehydrated eggs and potatoes. In a single day, the troops also consumed 2,500 loaves of bread and hundreds of pounds of butter. Though the Wakefield could convert salt water to fresh for bathing, the demand was so high that the troops sometimes showered with saltwater and Hershey's soap. Latrines were just a series of outhouses set over a metal trough that drained into the sea. Twice a day the whole deck was hosed-down due to seasickness.[161]

As with the troopship coming over, bunks were stacked in fours and packed in tight below decks. Rudolph Lea describes how his bunkmates made that work:

> The vertical four of us figured out the best way to make room was to place all four duffel bags on the floor under the bottom bunk somehow. That helped a lot. Then I put my musette bag at the head of the bunk as a pillow, my backpack at the foot, extricated my G.I. blanket sleeping bag from the duffel, climbed up, took off my G.I. clodhoppers and hung them with their laces from the ropes of the bunk above me, and lay down on top of the blanket bag. Other guys did pretty much the same, milling around below, going to the latrine, shuffling their gear, and trying to get settled.[162]

Despite ridiculously tight quarters, the troops were in high spirits, playing cards, reading, smoking, gambling, playing music and even jitterbug

---

[161] https://www.mycg.uscg.mil/News/Article/2447098/the-lone-blue-line-Uss-wakefieldthe-coast-guards-b-l-ferry-75-years-ago.

[162] *Topsy's GI Journey*, p. 145.

dancing. Unlike the zig-zagging eastbound trip to North Africa two years prior, the Wakefield[163] plowed straight ahead. No one feared attack, as all the German U-boats had been sunk and the Luftwaffe's fighter planes, like the rest of the Nazi war apparatus, had been stilled. They enjoyed a calm crossing in late spring. Many of the men chose to sleep on deck under the immense wash of the Milky Way, rather than below decks in those crowded bunks.

---

[163] Having transported nearly a quarter of a million soldiers in World War II, the Wakefield never sailed again. The great ship, as the SS Manhattan a flagship of the Cunard Luxury Line, was decommissioned in 1946, and finally scrapped in 1964.

# July 1945

The trans-Atlantic voyage home took seven days. At dawn on July 6th, everyone on deck shouted, "Land Ho!" as a distant gray line appeared on the horizon. The captain announced by loudspeaker that all were to pack up, clean their quarters for inspection, and get ready for the ship to dock in Boston by late afternoon. As they came into the harbor, pleasure boats, tugs and yachts appeared, horns blowing, flags waving, and the people onboard shouting "Welcome Home!" Then as a trio of tugboats pulled the great ship up to the dock, Rudolph Lea recalled:

> Crowds of people on the pier waved Stars and Stripes of all sizes.... Now a band began playing patriotic music and at one point, all of the boats blew their horns and whistles in one long deafening unison blast of welcome home, drowning out the music on the pier. The Wakefield answered with its own claxon blast of thanks.[164]

A convoy of school buses painted dull Army green stood waiting for the 7,000 soldiers onboard the Wakefield to disembark, driving everyone through a cheering city to Camp Myles Standish, where the men of the 99th

---

[164] *Topsy's GI Journey*, p. 156.

CMB had to hurry up and wait to get processed for 30-day furloughs ahead of deployment to Japan. In the chow line, they were served by German prisoners of war, who promised that this would be the best meal they'd had in years. Rudolph Lea recalled that they were right; the prisoners had prepared delicious roast beef, home-fried potatoes, string beans, and apple pie with vanilla ice cream. The troops retired with full bellies to new, all-white, spanking clean barracks, awaiting their furloughs and tickets home.[165] Lyn's came through by the following afternoon, and he caught a train for Richmond, VA. The folks back home knew he was somewhere between France and Fluvanna, but no one knew when he might arrive. Here I have to steal a passage from my mother's memoir:

> July 9th, 1945 dawned like any other sultry summer day. I walked the mile to the post office and back and had gotten ready for bed. After blowing out the oil lamp, I settled in for a hot summer night, since without electricity, our house had no fans. The windows were wide open to catch any whiff of breeze. All of a sudden came a shrill "bob white" call from down the road, just like in the first days of our marriage. I knew this was no quail. It was Lyn! There was no doubt in my mind. This was the sound I had dreamed of hearing for so many months and years. There were times when I thought I'd never hear it again.
>
> Rushing down the long flight of steps, wearing mismatched pajamas with curlers in my hair, I threw the door open. There he stood with a grin on his face, looking wonderful. But I knew this was not the same thin blond boy, little more than a teenager, who had left years ago, but a man who had slept with the stench of death in his nostrils, a soldier fresh out of combat, a survivor of a war that had taken him from this very door across the Atlantic Ocean to battle in North Africa, Italy, France and Germany itself. He had learned what war and death were all

---

[165] *Topsy's GI Journey*, p. 158.

about. He knew how it felt to have a friend fall at his feet never to rise again.[166]

They had a month's long reunion, which Mama recalled as idyllic. She wrote: "We spent every free minute together, and one weekend drove to the Shenandoah Valley, enjoying the beauty of the green mountains and visiting the Natural Bridge – then a big tourist attraction – for the first time." Here's how I imagine that month: They visited with Mama's brother Hollis, recuperating from a wound suffered as a tank crewman in Patton's 3rd Army during the Battle of the Bulge. He was living with his parents back on the farm and, as it turned out, spent the rest of his life at home with Grandma. Lyn's two brothers, Jimmy and Mac, were home, too, and they shared an outdoor family reunion picnic, all glad to have made it back, at least for a little while. Mama's older sister Nellie baked her delicious trio of pies: pecan, chocolate chess and lemon chess, and Lyn's sister Dorothy brought her famous caramel cake. Another day, they drove over to the Bethel Baptist Church cemetery to place flowers on the gravesite of Lyn's cousin Curtis, killed in the war.

Lyn's brother Jack was married to Virginia's sister Nellie. The two couples lived in an old two-story house with a wide porch that the Fork Union Military Academy rented to Jack, who was still milking a herd of Guernsey's twice a day to feed the cadets. Lyn went with him once or twice, for old time's sake, but most days, while he waited for his bride to come home from work, he carried a fishing pole through the woods behind the

---

[166] Mama worked at the post office during the war. As previously noted, she named her memoir *Then the Bob White Called*, to honor this, her most precious memory. You can buy the book on Amazon: https://amzn.to/3WNOjyj, or order it at your favorite bookstore.

house down a well-trodden path to the banks of the James River, where he cast a chicken gut-entwined bobber hook for catfish. Cicadas and crickets made the woods whine all day, out of tune with the roar in his head, all that shelling having spawned a tinnitus that would never go away, a constant, nagging memento of combat.

The river behind Fork Union ran shallow and fast; the catfish, some as long as his arm, lounged in the quieter rock pools. Lyn's sergeant had warned of what he called the heebie-jeebies -- the trembling and the racing heart and the tears. To get past it, he'd take off his boots, roll up his slacks, and wade into the water. Forcing a focused minute of here and now, he'd pause, eyes searching the glistening surface, then thrust down to wrest a catfish from its lair -- a writhing finned and mustachioed muscle -- with his bare hands.

Jack and Nellie had a tabletop radio, and Virginia brought the newspaper home from work every day. The headlines claimed that the Japanese were backed into a corner, but they seemed to be fighting with a suicidal intensity, even more fanatically than the Germans, giving up one Pacific Island after another, but at a huge cost to Allied troops. It was a different kind of conflict than Lyn and his Battalion were used to, fought on rugged beachheads and in dense tropical jungle, but there was nothing for it. The fighting in Europe may have ended but the Pacific War still raged.

*Pvt. Lyn Gentry – briefly back home in Fork Union, VA, July 1945*

# August 1945

On August 6th, recounts Mama's memoir, "a sad couple boarded the bus in Fork Union…headed for Broad Street Station in Richmond, where Lyn would leave for Camp Chaffee. Walking down the street for a sandwich before the train arrived, these words ran across a teletype: HIROSHIMA BOMBED BY ATOMIC BOMB: THOUSANDS KILLED." [167] She added, "This was horrible, but as we looked into each other's tear-filled eyes we could not believe the timing of this dramatic development that would change the course of history and hopefully end the war before Lyn had to go off to fight again."

The four-day trip out to Arkansas required two changes of train, but unlike the dingy, crowded troop trains Lyn had ridden at the beginning of the war, these were real passenger trains, with comfy seats, cloth doilies on the headrests, and porters in white gloves passing down the aisles serving Coca-Colas. Stretching his legs on the platform in Chattanooga, Lyn heard the newsboys shouting that another Japanese city, a place called Nagasaki, had just been struck by an even bigger atomic bomb than the one that had flattened Hiroshima three days before. This was the long-rumored weapon

---

[167] *Then the Bob White Called*, p. 61.

that would end the war, the one they'd all feared the Germans were developing. It was clear to all that if the Japanese continued to fight, President Truman – who had already approved devastating fire bombings of Tokyo and other Japanese cities – would use it again.

Camp Chaffee, situated near the Arkansas River midway between Little Rock and Tulsa, Oklahoma, in the hilly Ozark region of Arkansas, had been thrown together in just a year at the beginning of World War II as a training center for the Army. One section had been converted into a prisoner of war camp that held 23,000 POWs from the North Africa campaign, who were being processed for return home to Germany when Lyn arrived.

Gradually, the men of the 99th Chemical Mortar Battalion filtered into camp, and by the 26th, enough of the unit had checked in to begin redeployment training. But then some good news arrived. The Army had revised its thinking about the points system. Their officers announced that anyone with 80 points – medal winners with children -- would be discharged immediately. And anyone with 45 points could be assured that they would not be shipped overseas. This included Lyn and all of the men who had served with him since Texas. Only the replacement troops, who'd joined the Battalion later in the war, would go to Japan.

At the end of the day, more good news, when they assembled on the parade grounds to hear their commanding officer read the Presidential Unit Citation awarded for service with the 3rd Infantry Battalion during the Colmar Pocket campaign (see Appendix C).

Redeployment training began the next day, the old-timers who were going home almost nostalgically setting up and firing their mortars, showing the replacements all the tricks of the trade they hadn't had time to teach them on the battlefield: How to angle the tube accurately, how to gauge the extra distance a donut charge would add to a launch, how to bend their knees like a baseball batter in order to stand-and-pivot smoothly for the day-long passing along of an endless line of projectiles, how to keep your mouth open so incoming shells wouldn't blow out your ear drums, how to cuss correctly....

# September 1945

And then it was over. On September 2nd, the Japanese surrendered to end World War II. It took another month and a half for the Army to cut Lyn and his buddies loose; but on October 27th, he won his discharge. By then Virginia had taken the train out to join him, and they celebrated with another Army couple (the Vogt's – Private Richard Vogt had signed his swastika flag) in the resort town of Hot Springs before heading home again. By that time, Lyn had served three days short of three years at five Army bases in the States, in North Africa, Italy, France and Germany, and faced almost daily frontline combat for 150 days of that time.

Some other numbers. From Rick Atkinson's *The Guns at Last Light*:

By the time Japan surrendered on September 2,1945, the Second World War had lasted six years and a day, ensnaring almost sixty nations, plus sundry colonial and imperial territories. Sixty million had died in those six years, including nearly 10 million in Germany and Japan, and more than twice that number in the Soviet Union – roughly 26 million, one third of them soldiers. To describe this "great and terrible epoch," as George Marshall called it, new words would be required, like "genocide"; and old words would assume new usages: "Holocaust."[168]

Those are mind-numbing numbers, no doubt even to the men and women who survived the carnage. Probably, for each of them, what haunted the rest of their lives was the more personal losses, the ones that left their own hands bloody. I believe that for my father, it was that French farmer crumpled on the side of the road, his best friend lying ridiculously headless beside his mortar, nameless bodies in the rubble of mortar-pounded villages, stacks of burned and starved bodies near Dachau, and perhaps a death or two he found too difficult to ever speak of.

There was another side to the war, though, one that I've heard veterans mention many times. As one air force crewman, quoted by Atkinson, recalled, "Never did I feel so much alive. Never did the earth and all the surroundings look so bright and sharp." A combat engineer added, "What we had together was something awfully damned good, something I don't think we'll ever have again as long as we live."[169] Like soldiers in every war, the men of the 99th CMB fought for the men in the foxhole beside them, their trusted friends and brothers. What, in civilian life, would ever compare with the camaraderie they had shared?

---

[168] Atkinson, Rick. *The Guns at Last Light*, p. 632.

[169] Atkinson, p. 640.

# Afterword

"No war is really over until the last veteran is dead."[170]

The 99th Chemical Mortar Battalion of the United States Army was dissolved at the end of 1945. Seeing that coming, the officers of the Battalion took some time at Fort Chaffee to tell their story. That narrative, now online (https://www.4point2.org/hist-99.htm) is the backbone of this book; please take the time to read it. As the officers wrote in their prologue: "Here is mud, cold, and fatigue; here is heroism and plain courage; here are men who know danger and death, and who knew the satisfaction of a job well done. Ours was the task of supporting the "dough boy" and we are intensely proud of our accomplishments."

Another recommended site is a video collage of the 99th CMB's progress through the war, available at:

http://collections.ushmm.org/search/catalog/irn1000551.

---

[170] Atkinson, quoting an WWII infantry rifleman, p. 641.

Filmed by Battalion C.O. Lt. Commander Gordon Dixon, the spliced together footage is choppy, grainy, without narration. It plays like a 12-minute dream, its snippets tagging instants of the Battalion's wartime journey, from the sands of the Sahara to the horrors discovered at the Kaufering work camp. I've watched it over and over while writing this book, and every time it has given me a chill.

16 million Americans wore military uniforms during World War II. By 2024, it is estimated that just a hundred thousand will still be alive. My great regret in writing this book is that I waited too long. I never dared to skirt Mama's rule. I didn't consider reaching out to Daddy's platoon mates from Bluefield and Saginaw and Prescott and elsewhere who'd signed his flag until they were all gone (the last of them, I'm afraid, passed just last year: Corporal Byron D. Lemmon of Pocatello, Idaho (age 98), who'd been awarded two Purple Hearts and a Bronze Star). Fortunately, World War II has been exhaustively documented, so it has been possible to piece at least this much together. My hope is that some descendant of one of the guys who signed that flag may come across this book and contact me to share what their dad or granddad told them. And if they do, I'll add it into a later edition. But there's so much that I fear we'll never know about the 99th Chemical Mortar Battalion and Daddy's time with it.

Rudolph Lea wrote that the Battalion held occasional reunions over the years, but I don't recall ever hearing of more than one. That was a 50th anniversary tour of their path through France and Germany. Mama ends her memoir with excitement over the prospect of going on that trip, which would have taken them into Colmar and some of the other Alsatian battlefield villages, then on to Wurzburg, Neustadt, Munich and the

Bavarian Alps. We kids offered to help pay for the trip, but they never went and never said why.

My father was a gentle, haunted man. He worked at a string of blue collar jobs into his 70s and died, a few years after Mama, at the age of 81. Unlike the other men in our community, he didn't hunt, saying that he'd hauled a rifle in the woods all he cared to, thank you. He lived by a pair of maxims, often repeated in response to my callow whining: (1) If you can't say anything nice about somebody, don't say anything at all, and (2) the people in hell want iced water, too.

Daddy was a man of few words who drank alone. On occasion, iced water.

# Appendices:

## Appendix A:

Honorable Discharge Certificate

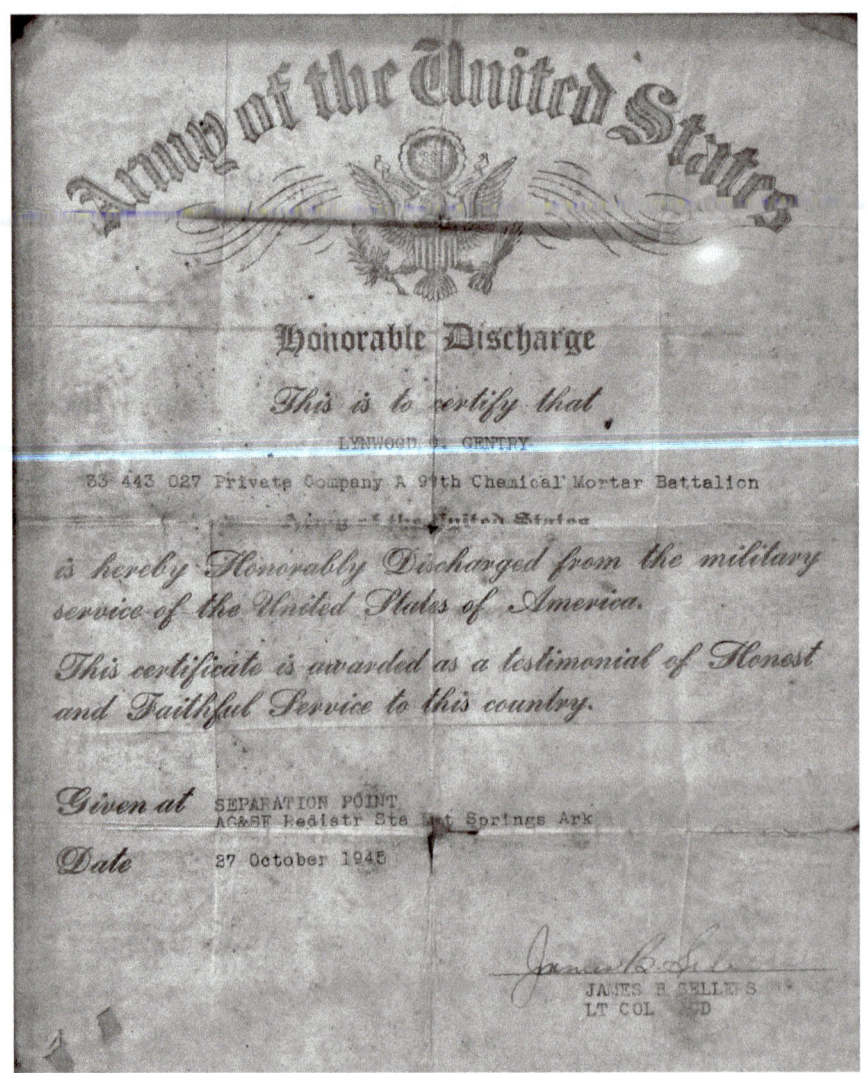

## Appendix B:

### Signatures on the Swastika Flag, signed March 26, 1945
### Service Awards Added

| | | |
|---|---|---|
| Staff Sgt. John Miksch<br>Weimar, Texas<br>Purple Heart<br>Soldiers Medal<br>Bronze Star | Staff Sgt. Evert Nelson<br>Turon, Kansas<br>Bronze Star | Sgt. Arlie Boatright, Jr.<br>R. Rt. 5<br>McAlester, Oklahoma<br>Bronze Star |
| Tech. 5 James E. Cooper<br>North Vernon, Indiana<br>Purple Heart<br>Bronze Star | Cpl. Byron D. Lemmon<br>Pocatello, Idaho<br>Purple Heart with Oak Leaf Cluster<br>Bronze Star | Cpl. John J. Richards or Don Pickerel*<br>Ust, Washington*<br>If Richards –<br>Bronze Star |
| Pfc. E. Anderson, Jr.<br>Cheney, Wash<br>Bronze Star | Pfc. Basil K. Morehead<br>4671 W. 150 St.*<br>Cleveland, Ohio<br>Purple Heart | Pfc. Vernon K. Perry<br>Prescott, Arizona<br>Purple Heart<br>Silver Star<br>Bronze Star |
| Joseph Denson<br>1208 Lamben St.<br>Galveston, Texas*<br>Bronze Star | James Beggs<br>3331 NW Sauer St.*<br>Portland, Oregon | Pfc. Andrew W. Castando*<br>Casa Grande, Ariz. |
| Joseph Dressler*<br>1208 Lamben St.<br>Galveston, TX | Pvt. Charles Engbert *<br>4358 N. Elston Ave.*<br>Chicago, Il. | William Filan<br>Lawton, Penna. |
| William Goldstein<br>Van Nuys, Calif. | Carter Howard<br>Cottonburg, Ky. | Pfc. J. G. Infante<br>P.O. Box 1590<br>McAllen, Texas |
| Ellwood Landt<br>Espanola, Washington | Mike C. Magnotta<br>206 N. 62nd Street<br>Philadelphia, PA | Thurman L. Norton<br>Mascot, Tennessee* |

| James Nowlin<br>Bluefield, W. Va. | Dewey A. Oliver<br>Memphis, Tenn | Sigfried F. Sandos<br>Gardner, Massachusetts |
|---|---|---|
| Leo Thibault<br>Fall River, Mass. | Richard Vogt<br>2348 So. Michigan Ave<br>Saginaw, Michigan | Pvt. Guy W. Wela<br>Rockford, Illinois |
| J. Zuke<br>Morris Run, PA | | |

*My best guess at illegible ink

Appendix C:

Extract of Unit Citation[171]

WAR DEPARTMENT
GENERAL ORDERS No. 44
Washington 25, D. C., 6 June 1945

EXTRACT

BATTLE HONORS - Citations of Units

XIII. BATTLE HONORS
1. [omitted]
2. As authorized by Executive Order 9396 (sec. I, WD Bul. 22, 1945), superseding Executive Order 9075 (sec. III, WD Bul. 11, 1942), the following unit is cited by the War Department for outstanding performance of duty in action during the period indicated, under the provisions of section IV, WD Circular 333, 1943, in the name of the President of the United States as public evidence of deserved honor and distinction. The citation reads as follows:

The 3rd Infantry Division with the following attached units:
254th Infantry Regiment
99th Chemical Battalion
168th Chemical Smoke Generator Company
441st Antiaircraft Artillery Automatic Weapons Battalion
601st Tank Destroyer Battalion (SP)
756th Tank Battalion
IPW Team 183

fighting incessantly, from 22 January to 6 February 1945, in heavy snow storms through enemy-infested marshes and woods, and over a flat plain crisscrossed by numerous small canals, irrigation ditches, and unfordable streams, terrain ideally suited to the defense, breached the German defense wall on the northern perimeter of the Molmar bridgehead and drove forward to isolate Colmar from the Rhine. Crossing the Fect River

---

[171] https://www.4point2.org/hist-99.htm.

from Guemar, Alsace, by stealth during the late hours of darkness of 22 January, the assault elements fought their way forward against mounting resistance. Reaching the Ill River, a bridge was thrown across but collapsed before armor could pass to the support of two battalions of the 30th Infantry on the far side. Isolated and attacked by a full German Panzer brigade, outnumbered and outgunned, these valiant troops were forced back yard by yard. Wave after wave of armor and infantry was hurled against them but despite hopeless odds the regiment held tenaciously to its bridgehead. Driving forward in knee-deep snow, which masked acres of densely sown mines, the 3d Infantry Division fought from house to house and street to street in the fortress towns of the Alsatian Plain. Under furious concentrations of supporting fire, assault troops crossed the Colmar Canal in rubber boats during the night of 29 January. Driving relentlessly forward, six towns were captured within 8 hours, 500 casualties inflicted on the enemy during the day, and large quantities of booty seized. Slashing through to the Rhone-Rhine Canal, the garrison at Colmar was cut off and the fall of the city assured. Shifting the direction of attack, the division moved south between the Rhone-Rhine Canal and the Rhine toward Neuf Brisach and the Brisach Bridge. Synchronizing the attacks, the bridge was seized and Neuf Brisach captured by crossing the protecting moat and scaling the medieval walls by ladder. In one of the hardest fought and bloodiest campaigns of the war, the 3d Infantry Division annihilated three enemy divisions, partially destroyed three others, captured over 4,000 prisoners, and inflicted more than 7,500 casualties on the enemy.

BY ORDER OF THE SECRETARY OF WAR:
G. C. MARSHALL
Chief of Staff

OFFICIAL:
J. A. ULIO
Major General
The Adjutant General

## Appendix D:

Mama's memoir includes a recipe for snow ice cream, so thought it appropriate to include a dessert recipe here, too.

### D Ration Fudge

Ingredients:

1 block D ration chocolate
1 packet sugar
1 can condensed milk
1 large shell fragment

Directions:

Pound chocolate with shell fragment until powdered. Throw away shell fragment. Mix chocolate with sugar and milk over Coleman stove. Test by dropping samples of mixture in canteen cup of cold water. When sample congeals in water, pour fudge into shallow pan, cool and slice.[172]

---

[172] From Kennett, L. *The American Soldier in WWII*, p. 102.

# Bibliography

Ambrose, Stephen. *Band of Brothers*. New York: Simon & Schuster, 1992.

Ambrose, Stephen. *Citizen Soldiers*. New York: Simon & Schuster, 1997.

Atkinson, Rick. *An Army at Dawn*. New York: Henry Holt, 2002.

Atkinson, Rick. *The Guns at Last Light*. New York: Henry Holt, 2013.

Brooks, E. Kleber. *The Chemical Warfare Service: Chemicals in combat*. Washington, DC: Center of Military History, United States Army, 1990.

Champagne, Dan. *Bloody Fight for Hill 351: Skirmish in the Colmar Pocket*. Warfare History Network: Fall 2021. https//warfarehistorynetwork.com/article/bloody-fight-for-hill-351/.

Clarke, Jeffrey J. & Smith, Robert Ross. *Riviera to the Rhine*. Washington, DC: U.S. Government Printing Office, 1993.

Daly, Hugh C. *42nd "Rainbow" Infantry Division: a combat history of World War II*. World War Regimental Histories. United States Army, 64, 1950. http://digicom.bpl.lib.me.us/ww_reg_his/6.

Eldredge, Walter J. *Finding My Father's War: A Baby Boomer and the 2nd Chemical Mortar Battalion in World War II*. Otsego, MI: PageFree, 2002.

Gellhorn, Martha. *The Face of War*. New York: Grove Press, 1998.

Gentry, Virginia G. *Then the Bob White Called*. Richmond, VA: Nextext, 2021.

Hastings, Max. *Armageddon: The Battle for Germany 1944-1945*. New York: Knopf, 2004.

Hoyt, Edwin P. *The GI's War: The Story of American Soldiers in Europe in World War II*. New York: McGraw-Hill, 1988.

Kennett, Lee. *G.I.: The American Soldier in World War II*. New York: Scribner's, 1987.

Kershaw, Alex. *Against All Odds: A True Story of Ultimate Courage and Survival in World War II*. New York: Dutton Caliber, 2022.

Kotlowitz, Robert. *Before Their Time: A Memoir*. New York: Knopf, 1997.

Lea, Rudolph. *Topsy's GI Journey: Tales of a Soldier and his Dog in WWII*. New York: iUniverse, 2007.

MacDonald, Charles B. *The Siegfried Line Campaign*. United States Army: Center of Military History, 1993.

Miller, Katherine I. *War Makes Men of Boys: A Soldier's World War II*. College Station: Texas A&M University Press, 2013.

Miskimon, Christopher. *Destruction of the Colmar Pocket*. Warfare History Network, Winter 2023. https://warfarehistorynetwork.com/article/destruction-of-the-colmar-pocket/

Murphy, Audie. *To Hell and Back*. New York: Grosset & Dunlap, 1949.

Norris, John. *Mortars in World War II*. Barnsley, England: Pen & Sword Military, 2015.

The Officers of the 99th Chemical Mortar Battalion. *Our Part: A History of the 99th Chemical Mortar Battalion During Combat in Africa, Italy, France, and Germany*. Camp Chaffee, Ark., September 1945. https://www.4point2.org/hist-99.htm

Phibbs, Brendan. *Our War for the World: A memoir of life and death on the front lines in WWII*. Guilford, CT: Lyons, 2002.

Prefer, Nathan N. *Eisenhower's Thorn on the Rhine: The Battles for the Colmar Pocket, 1944-45*. Philadelphia: CaseMate, 2015.

Roberts, Andrew. *The Storm of War*. New York: HarperCollins, 2011.

Thompson, Clinton W. *Hell in the Snow: The U.S. Army in the Colmar Pocket, January 22-February 9, 1945*. University of Texas at Tyler, 2017.

https://scholarworks.uttyler.edu/cgi/viewcontent.cgi?article=1011&context=history_grad

Vannoy, Allyn. *Delaying Action at Enchenberg*. Warfare History Network, August 2010.
https://warfarehistorynetwork.com/article/bloody-fight-for-hill-351/

Wheeler, James Scott. *Jacob L. Devers: A General's Life*. Lexington, KY: University Press of Kentucky, 2015.

ABOUT THE AUTHOR

Tony Gentry is a writer and professor emeritus in occupational therapy at Virginia Commonwealth University. Much of his career in practice and research has involved military veterans. He lives in Bon Air, Virginia with his wife Christine, a polytrauma OT at Richmond's Veterans Affairs hospital.

Website: https://www.tonygentry.com

Email: tonygentrywrites@me.com.

www.ingramcontent.com/pod-product-compliance
Lightning Source LLC
Chambersburg PA
CBHW060520100426
42743CB00009B/1390